Merry Keating

How to Assure Quality in Plastics

Hanser Publishers, Munich Vienna New York

Hanser/Gardner Publications, Inc., Cincinnati

The Author:
Merry Keating, Superior Plastic, Inc., 417 E. Second Street, Rochester, MI 48307-2007, USA

Distributed in the USA and in Canada by
Hanser/Gardner Publications, Inc.
6600 Clough Pike, Cincinnati, Ohio 45244-4090, USA
Fax: (513) 527-8950
Phone: (513) 527-8977 or (800) 950-8977

Distributed in all other countries by
Carl Hanser Verlag
Postfach 86 04 20, 81631 München, Germany
Fax: +49 (89) 98 12 64

The use of general descriptive names, trademarks, etc., in this publication, even if the former are not especially identified, is not to be taken as a sign that such names, as understood by the Trade Marks and Merchandise Marks Act, may accordingly be used freely by anyone.

While the advice and information in this book are believed to be true and accurate at the date of going to press, neither the authors nor the editors nor the publishers can accept any legal responsibility for any errors or omissions that may be made. The publisher makes no warranty, express or implied, with respect to the material contained herein.

Library of Congress Cataloging-in-Publication Data
Keating, Merry.
How to assure quality in plastics / Merry Keating.
    p.   cm.
Includes bibliographical references and index.
ISBN 1-56990-180-5
1. Plastics—Quality control.   I. Title.
TP1122.K43   1995
668.4'068'5—dc20   95-2599

Die Deutsche Bibliothek - CIP-Einheitsaufnahme
Keating, Merry:
How to assure quality in plastics / Merry Keating. - Munich ;
Vienna ; New York : Hanser ; Cincinnati : Hanser/Gardner,
1995
    ISBN 3-446-17516-4

© Carl Hanser Verlag, Munich Vienna New York, 1995
Typeset in the USA by pageAbility, Watertown, NY
Printed and bound in Germany by Kösel, Kempten

Preface

In today's highly competitive plastics industry, the integration of a strong and complete Quality Program is essential. There are many books and training programs available that detail the various technical disciplines (tools) commonly used within a successful Quality Program. This book will aid any individuals involved in Quality or Management who are interested in using those tools to implement, maintain, and continuously improve an effective Quality System in any plastics business. Case studies and examples will be included that highlight typical roadblocks and the methods followed to overcome them. Throughout the book the "team" concept will be defined in detail along with the pragmatic steps involved in team building. It will take you beyond the orthodox boundaries of "quality" and address relationships, roles, and the art of follow-up which truly are essentials in the quest for quality.

This book (for example) will *not* teach you the formulas to calculate "analysis of variance" for a design of experiment, but will address a rather practical approach, once obtaining education, on using the tools to create a process optimization program. The intention of this book is to give the reader a basic understanding of each of the areas discussed. However, the focus will be on how to use the tools effectively to gain and maintain beneficial results and continuously improve the results.

My appreciation goes out to John Rake, of C.R.B. Crane, for truly empowering me in my roles throughout the years. I thank Jackie Shea, my friend and a Quality Liaison for American Yazaki Corporation, who took the time to edit the text and language for me on the manuscript. Lance Neward of Perrin Manufacturing and past President of the S.P.E. should also be acknowledged for (without realizing it) helping me keep my focus on "how to make the tools WORK". Rocky Kimball of Lacks Enterprises, a role-model from the past, I thank for reviewing the chapter on ISO-9000. Finally, I thank Norman Mackie, President of Superior Plastic, Inc., for the overall support I received from the company during my work on this project.

Merry Keating

Contents

1   The Beginning

1.1   Quality Policy

Like any other department within an organization, the quality assurance function has a *purpose*. The foundation of a successfully implemented quality system is the "Quality Policy." I use the term "policy" here, as it has been referred to as that for many years within the production industry. To better name the "purpose" however, I choose to refer to it as the "mission."

The documented mission will ensure consistency of the purpose, establish a philosophy, and provide the focal point in which individuals can identify. Basically, it will become the translation of all the quality related procedures and programs to follow.

Because the quality department's mission must be congruent with all company objectives, it is essential that top or corporate plant management is involved in developing the policy. In addition to plant management, the entire Quality Department and representatives from all levels of the organizational structure from every department should be encouraged to participate.

Someone should be appointed by the group to facilitate when the initial meeting takes place. It has always been my choice to suggest an individual from the base of the organizational structure to facilitate ("base" personnel are not salaried or supervisory personnel). This would imply immediately to the group that no one person is more important than another for the purpose of this particular meeting and what is to be accomplished.

Once the "facilitator" has been selected, he/she should be given the following guidelines to conduct the meeting:

- Tell all participants there are no wrong answers or suggestions.
- Keep everyone from talking at once (go around the table one at a time if you must), and encourage everyone to participate.
- Ask these questions:
  Who are our customers?

What service/product do we provide?

What can we commit to in regard to quality in satisfying our customers?

What are our concerns for growth, survival and profitability?

Where do we compete and what is our distinct advantage over the competition?

- Record all of the replies to the following questions:
  What is our mission in regard to quality?
  How will we accomplish that?
  Can you define that?
- Organize all recorded notes and make copies for each member in the meeting.
- Write the mission together. Every component of the mission/ policy must be agreed upon by all participants.
- When the mission is finalized, it should by typed and then signed by every member of the team.
- All team members should be given a copy of the signed mission and a copy should be placed in the front of each department's plan and/or procedures manual.

Remember, a Quality Policy written by one or two people is the mission of only one or two people. To have complete "buy in" to the mission, you must have complete investment by everyone who must *believe* in it.

There is no such thing as a "poor" Quality Policy; the effect of the policy is what measures its worth. Included here is an example of one of my favorite policies. Keep in mind, it is one I prefer because it was affirmed by all members of the company.

Example

We are dedicated to the principle of "never-ending improvement in quality and productivity." Our goal is to be innovative in quality and technology. We will make quality-related decisions as if the product produced was destined for our own use. We will achieve this goal through control of our processes with the aid of statistical techniques. The process is defined as operators, materials, machines, environment, and methods. Every employee is fully committed to this philosophy in the performance of his/her duties.

Your entire Quality Policy may consist of only one sentence. Imagine that you are the customer visiting one of your supplier's manufacturing facilities.

Imagine how you would feel if this supplier had a Quality Policy that simply read: "Uncompromising service to our customers," and all employees truly *lived* this mission. I believe you would leave that facility as a satisfied customer, confident of the quality product they will send to your company.

1.2   Personnel Organization and Job Descriptions

The size and organizational structure of a Quality Department will depend mainly on the size, output, and process complexity of the manufacturing area. There is also typical variation from one company to another in regard to certain quality-related functions that may be located within other departments. To operate along the guidelines of this book, the Quality Department should represent approximately 4 to 6 percent of the entire work force.

More often than not, our customers require that the Quality Function be separate from manufacturing and ultimately report to a separate management. I think the reasoning behind this type of customer-driven requirement stems from past paradigms that may have made sense in the past. It was not so long ago (and still current in many companies) that product inspectors and technicians were considered to be the "police," "bad guys," "nags," "nit-pickers," etc., by the manufacturing personnel. Quite frankly, it was thought that if the quality function was not a segregated operation that Production Control, Engineering, and Manufacturing Departments would merely consider their efforts negligible and totally disregard them. This example is somewhat severe, but real.

Only companies that have top/corporate management enforcing the philosophies of "quality" and "continuous improvement" can unite the quality and manufacturing functions. I should also add here that for this union to be effective, manufacturing should be held responsible for quality. By "held responsible," I suggest to the point that *manufacturing* is perceived to control and have *ownership* in the quality being manufactured and shipped.

The structure of the quality department will vary from company to company depending on the types of job descriptions within that department. In any case, you should never have any individual reporting directly to two or more persons or have everyone in the department on the same level and reporting to the top manager. I will explain the reason for this statement, even though I would assume the reasons are inherently evident.

When an employee has more than one "boss," it is very difficult for him/her to prioritize assignments. It is also impossible for either leader to have a clear idea of the employee's workload and can easily create a situation of animosity between any combination of all persons involved. It is acceptable for an employee to have more than one resource of expertise in regard to different job duties if structured correctly. On an organizational chart, you would depict those superiors/resources at a higher level but have them connected to the employee by a dotted, rather than a solid line. Those responsible for an employee connected by a dotted line represent communication, information, and guidance but would not assign duties or manipulate the employee's workload without consent of the direct supervisor.

A department in which all or most of the employees report to a single person will experience and cause problems, unless the department is very small (3–4 job titles with no more than 3 of each title). This type of structure lacks empowerment of the employees and introduces potential for stagnation in the department's performance and innovative growth. Its capabilities will be limited to the individual capabilities of the leader. In addition to these limitations, problems will arise if the manager is not available for the company for whatever reason or if he/she becomes no longer employed at the company. This flat organizational structure would make it difficult to know who to put in charge until a replacement was found or the manager returned.

A reporting structure should be defined and branched out to insure maximum empowerment and capability potential of all individuals, (reference Exhibit 1). It should also be structured as such to indicate each individual's path to promotion.

1.2.1   Job Descriptions

There should be a formal and documented job description for every position within the company. Each job description should contain, at a minimum, the following information:

- Job title and department name,
- supervisor to whom employee directly reports,
- effective date and last revision date,
- responsibilities and duties,
- education and experience required,
- all physical requirements (in accordance to the Americans with Disabilities Act), and
- an approval signature by a company officer and/or the Personnel or Human Resources Manager.

It is advantageous to include a statement on the job description that would cover unexpected or rare needs of the supervisor. The statement could read, "Performs other duties as instructed by the supervisor or as necessary to maintain efficient operations." Each employee should have a copy of his/her job description in their possession for reference. As well, the "master" of all job descriptions should be maintained by the Personnel or Human Resource Manager. I have found it to be convenient to include the Quality department's job descriptions in the Quality Manual also.

It is the direct supervisor's responsibility to ensure that each employee fully understands and has been trained to carry out the responsibilities and duties listed on his/her job description. It would be beneficial to have each employee sign or initial a statement that signifies he/she indeed has reviewed, understood, and received a copy of their job description. The signed statement could be entered into each individual's personnel file.

Revisions and modifications of job descriptions should occur as result of annual (at a minimum) reviews that reveal changes in the current description. As well, any other time that changes occur within the department that would affect the current job description, it should be updated. Any time a job description is revised/modified, it should be reviewed with the affected employee and an updated copy should be issued to him/her.

As I mentioned before, *all* titles within the department need a job description. I revisit this necessity because it is a very common error of Quality Assurance/Control Managers and Corporate Directors to neglect the fact that they, too, need a job description.

1.3   Quality Manual

The quality manual should contain or reference every quality-related procedure. If there are procedures specific to a certain area within quality, (i.e., reliability, etc.), you may choose to create a separate manual for that area. If this is the case, be sure to include a table of contents for that specific manual in the main quality manual along with a notation of where the separate one is located and who controls it.

Many customers require that every procedure in the manual be signed/approved by the top manager in the quality department. An easy way to meet this requirement is to do just that. The main thing to be accomplished is to ensure that each procedure is approved and considered "in effect." If signing each procedure is for some reason not practical or if you feel someone's signature "dates" the document, there are other ways to denote procedure approval. In some companies, the Controller or another activity maintains all of the company's procedures and they are considered "approved" once that person has assigned a "procedure serial number" or another type of I.D. number. When this type of procedure approval method is used, it is important to have the details of that process documented and included in the manual.

If you look at the table of contents for this book, you will get an idea of what the table of contents for a quality manual should look like. At a minimum, it should contain:

- The Quality Policy/Mission,
- the Quality Department's organizational chart,
- a revision record,
- a statement of who controls the manual and where the "Master Manual" is located,
- a list of all functions that have copies of the manual,
- quality-related procedures, and
- job descriptions for all quality related personnel (though typically kept in the Personnel Department, can also be included in the quality manual).

It is essential that only one person in the organization is in control of the "master manual." That person should also be the one who is responsible for assuring that a copy of the manual is available at all sites where the document is required for effective operations. I recommend that each copy of the manual is given a serial number and signed for by the person responsible for its maintenance at their job site. A record of the assigned numbers, names, and location will also make it easier to track the manuals when updates or audits occur.

An effective way to ensure that quality will be improved continuously is to hold a multi-disciplined meeting every quarter or twice a year to evaluate the results of your "Self-Assessments" (refer to the chapter in this book entitled "Internal Systems Control"). The results of your "in-house" assessments should enable the team to identify potential deficiencies in various procedures that led to less than perfect scores during the audit. In addition, a review of each procedure could lead to upgrading/clarification/improvement modifications needed for assorted reasons. In turn, you could then use the self-assessment scores to measure quality improvement. Please refer to the "Document Control" chapter to learn how to manage *revisions* to the manual and format *procedures*.

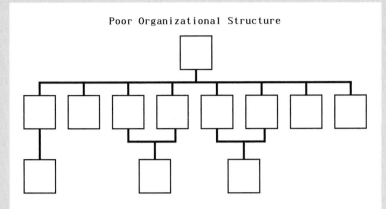

Poor Organizational Structure

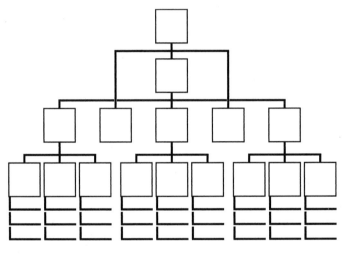

Good Organizational Structure

Exhibit 1: Organizational structure flow charts: Poor (top) and good (bottom)

2 Advanced Quality Planning

2.1 Voice of the Customer

Quality Function Deployment (QFD) is a methodology which serves as a tool that provides managers and/or planners with a translation of customer requirements into the company's guidelines for product development. Many companies have been creative in modifying the visual representation of the QFD tool for other types of planning and programs, though the original intention was for the translation of customer requirements for product development. The main reason for the use of QFD is to equip the manager or planner with a visual representation of *the voice of the customer.*

For the purpose of describing the benefits and process of QFD, let us assume here that "product development" is the area we intend to use it. We should find that committed utilization of the tool would result in fewer start-up problems, fewer engineering changes, and quicker product development phases. These improvements are bound to lead to a more satisfied customer which, in turn, will lead to increased market share.

So the purpose of QFD (in product development), is to ensure that the development of product features, characteristics and specifications, as well as the selection and development of process equipment, methods and controls are driven by the demands of the customer. It lends itself towards the prevention of problems rather than the detection of them and, therefore, reduces the overall amount of time spent in and after product development.

QFD is a tool intended for *team* use. The multi-disciplined team should consist at least of engineering, manufacturing, quality, production control, purchasing, and sales personnel. The template for QFD is the "house of quality." This is where the team will define and "house" all of the following information. (Please refer to Exhibit 2 for the location within the "house of quality" in which this information is stored.)

Once the House of Quality is constructed, the team can study the matrix, assessments, and ratings in order to understand where conflict exists. Conflict would be represented by a "weak" relationship symbol between a

Table 2.1 The QFD Process

A. Define the stages of product development:
    Marketing, Product Planning, Product Design, Engineering, Prototype, Testing, Process Development, Manufacturing, Assembly, Sales, Service

B. Define *what* the customer requirements are for each stage

C. Define *how* you will meet those customer requirements

D. Define *how much* (a measurable of *how*), which is how *how* can be accomplished

E. Assess the strength of *what's* relationship to its *how* by rating each relationship with a "weak," "medium," or "strong" symbol

F. Describe the correlation between each *how* item with the use of symbols that represent a "strong positive," "positive," "negative" and "strong negative" correlation. Note that you are only looking at correlations between the individual *how* items—*not* between the *how* and *how much* items.

G. Assess the competitive strength of each *what* item by using customer-oriented information. Rate your competitive strength as "high," "medium," or "low"

H. Assess the competitive strength of each *how* item by using engineering generated information. Again, rate each as either "high," "medium," or "low"

I. Rate the *what* items on a scale of 1–5 (5 being the strongest weight) to reflect the relative importance of each item to the customer

J. Multiply each *what* item's importance rating by a given weight of each *how* relationship strength and total those figures for every *how* column. Refer to the "E" step in this process, where you rated the strength of the relationship between *what* and *how* as "weak," "medium," or "strong" and now give each of those ratings a number, (i.e., "3," "6," and "9," respectively). (Refer to Exhibit 3 for an example.)

what and *how* item, a "negative" or "strong negative" correlation symbol between two or more individual *how* items, a low or medium competitive strength assessment (especially when in the area of any *what* item that has received a high importance rating). Once conflict is seen in the House of Quality, appropriate preventative actions can be taken to reduce and eventually eliminate those conflicts. As conflicts are reduce and eliminated, your company will begin to recognize the benefits of QFD.

2.2 Design and Feasibility Reviews

Design reviews are a series of reviews to ascertain whether a proposed design can be manufactured, assembled, tested, packaged, and shipped at acceptable levels and that production volumes and schedules are consistent with the ability to meet engineering, quality, reliability, investment, cost and timing objectives. They are typically initiated by the Project Engineer and occur in three phases. Phase 1 is the "design concept," phase 2 is the "prototype design" and "test," and phase 3 is the "final design." A design review is definitely multi-disciplined in each phase with the most overall involvement occurring in the second phase.

2.2.1   Phase 1

The Project Engineer conducts the design review and controls all documentation. The Design Engineer would prepare and present all of the design and data information to the team. The Design Engineer and Lab Analyst would evaluate the design for optimum reliability while the Sales Representative assures that all customer requirements are realistic and understood. Internal approval of phase 1 would come from the Engineering Manager who makes final evaluations of the design and assures that it adequately meets all of the customer's requirements. Overall final approval of the design concept will normally come from the customer who intends to purchase the product.

2.2.2   Phase 2

The prototype design review would be very similar to the phase 1 design concept review but with many more considerations. In addition to the same phase 1 activity, the Quality Manager would ensure that all inspection, control, and testing functions were established and conducted. The Manufacturing Manager would make sure that the design can be produced at the established cost and time goals and work with the Project Engineer to insure that all assembly, maintenance, and other considerations have been included. Purchasing would be responsible for assuring that required materials and components will be available at the established cost and time goals. The Project Manager and Lab Analyst should insure that the selected materials will perform to requirements, and the Packaging Engineer is to assure that the product can be handled without damage, etc. The tool source should also evaluate the design and affirm that cost and timing are on target and that tolerances and function requirements will be satisfied. At this point of the design review, the Sales Representative should not have to go back and assure that the customer requirements are realistic and understood. Fully carrying out phase 1 would have eliminated a need for revisiting that. Again, however, the customer should be involved to voice acceptability and/or make suggestions.

2.2.3   Phase 3

Review of the final design would progress exactly as phase 2, except for assuring that material/products perform and can be purchased at the time/cost goals. As well, the tool source should not be needed to re-evaluate the design.

Feasibility analysis, as demonstrated in the evaluations of the above phases of design review, should also occur after the final design (phase 3) and after the engineering release before the first prototype trial run. During feasibility reviews, analytical techniques such as Design of Experiments and Cause & Effect Diagrams should be utilized to optimize manufacturability and reliability. Quality Function Deployment should (QFD) be used to prevent any reasons to go "back to the drawing board."

Before the first trial run the following documents should have been optimized through all of the feasibility analysis stages and be available at the trial run:

- Process flow chart,
- process Failure Mode and Effects Analysis (FMEA),
- control plan,
- process instructions,
- process potential study,
- gage R&R, and
- packaging plans.

Final feasibility reviews should be scheduled after the trial run and the dimensional/functional integrity of the product validated. Refer to Exhibit 4 for a flow chart of this process.

2.3   Process Flow Charts

The Process Flow Diagram depicts with symbols, the "flow" of a process. It should be the first living document prepared for the manufacture of a product and will aid as a road map for the creation of a Process FMEA, Control Plan, and inspection instructions. The act of preparing a Process Flow Chart has, in some cases, even provoked product design changes and changes to the originally intended process/flow. The chart serves as a visual tool that makes it easier to detect segregated processes that potentially could overlap in order to eliminate subsequent operations. It might even make it suddenly clear that a design change or secondary operation fixture modification could significantly reduce the total cost of the finished product.

The intended process and its flow have to be known in order to give a cost quote to a customer or estimate a budget. It is from that perceived intention that the foundation of the Flow Chart is initiated and enhanced. To begin, let a symbol (refer to Exhibit 5) represent each type of stage in a process:

Symbol	Meaning
Open arrow	Movement
Square	Inspection/test
Triangle	Storage/hold
Diamond	Operation

Connect the symbols as they are intended to be connected in the actual process. Inside or beside each symbol, describe the stage of the process. The Process Flow Chart can then be elaborated on by also listing important product characteristics, how failures of those would be detected, and the process characteristics/ parameters that control each product characteristic (refer to Exhibit 6).

The Process Flow Chart should be placed on the manufacturing floor at each point of the process for reference. Because it is a living document, it should contain an Engineering Release Level, origination and last revision date, and an approval signature in addition to the product name and part number. The chart should be reviewed at established frequencies and whenever process or product changes occur. Continuous updates and modifications to the document will accommodate the representation of the current process.

2.4   Failure Mode and Effects Analysis (FMEA)

Failure mode and effects analysis is beneficial in recognizing and evaluating potential failure of a product, identifying actions that would potentially prevent the chance of failure occurrence, and documenting the chronological steps of the process being analyzed. It is intended for use before the *design* stage of a product and again before the production *process* is finalized. FMEAs are expected to have been constructed by a team consisting of members from engineering, manufacturing, quality, assembly, and sales/service. Refer to Exhibit 7 at the end of this section for an example of the flow for this process. The preparation of a FMEA could take place perhaps in a weekly Engineering meeting, or the documentation could be originated by one individual and circulated to the other functions for input to be returned to the initiator. I prefer an "all at once" group meeting because experience has demonstrated that overall time involved is minimized and the content more refined. Many suppliers of plastic products receive the product design from the customer and therefore would not initiate a Design FMEA, though they could participate in its development with the customer. The number of suppliers who design the products that they produce grows tremendously every day as many customers now expect suppliers to have design capabilities. Because this book is geared towards the Quality Function's tools, I will focus on the construction of a Process FMEA; although the concept is the same for both.

To begin the development of a Process FMEA, the form used should contain the following label information:

1. Part or process name and number,
2. Design origination and manufacturing location,
3. Any other plants, organizations, departments, etc. involved,
4. Suppliers,
5. A model year and customer project identification,
6. Engineering release date,
7. Production date,
8. Name of the individual that is responsible for the construction of the FMEA,
9. Project Engineer, and
10. Date of original FMEA date and last revision date.

All of this information should appear at the top of the FMEA. Refer to Exhibit 8 at the end of this section.

The rest of the form should be divided into vertical columns, each column containing the following information:

Column 1   List each *process description* (i.e., injection molding, blow molding, extrusion, sub assembly, heat-stake, etc.). This information can be taken directly off of the Process Flow Chart's "operations."

Column 2   List all *potential failures* for the product during each process. Surrogate data and past history knowledge of similar products/materials/processes is the largest source of information for predicting potential failures. A failure is defined here as any part of the product not meeting engineering specifications and/or anything the customer may consider objectionable. In considering who the "customer" is, always think in terms of who will buy the product directly from the company, who the final owner will be, as well as the next operations. Typical failure modes could be: flash, shorts, undersized hole, cracks, missing foam component, dirty, warped, splay, bubbles, discolored, mislabeled, strength, etc.

Column 3   List the *potential effect(s)* of each failure on the customer(s). Potential effects may be described as: noise, inoperative, excessive effort required, poor appearance, impaired operation, loose/unstable, will not fit, will not match, endangers operator, etc.

Column 4   The next column should be labeled as *Severity* and a *severity rating* applied next to each potential effect. Here a number will be used to represent the seriousness of the effects to the customer(s) on a scale of 1–10. Severity ratings are as follows:

Severity of Effect	*Rating*
Minor. Unreasonable to expect that this failure would cause a real effect. Customer would probably not even notice the effect.	1
Low. Nature of failure causes only slight deterioration or inconvenience.	2 3
Moderate: Failure causes some dissatisfaction or annoyance to the customer.	4 5 6
High: High degree of customer dissatisfaction that does not violate safety and/or government regulations. If the effect was labeled as being inoperable, causing serious disruptions or injury	7 8

to an operator, a high severity rating would be necessary.

Very high: Failure affects safe end-customer use of product and/or involves noncompliance with government regulations.
9
10

Column 5 List the *potential cause(s)* of each failure in terms of something that can be corrected or controlled. If many different causes or no known causes can be attributed to the reason for a failure, then further investigation (such as a design of experiment) should be utilized to determine which contributors can be easier corrected or controlled. Specific causes should be listed (i.e., tool worn or damaged, handling damage, inaccurate gaging by inspector, inadequate gating/venting, incorrect speed, hold pressure incorrect, filler content too high, etc.). Vague causes such as "operator error" and "malfunction" should not be used.

Column 6 This column should be labeled as *Occurrence* and an *occurrence rating* determined and applied for each potential failure. This rating does not measure the occurrences of customers receiving the failed product but rather the actual occurrence of the failure or cause of it. The probability of expected occurrence is rated on scale of 1–10. Occurrence ratings are as follows:

Occurrence	*Rating*
Remote: Failure not likely. Less than 1 in a million expected. Surrogate Cpk > 1.67 and is statistically in control.	1
Very low: Isolated failures. 1 in 20,000 is expected. Surrogate Cpk > 1.33 and is statistically in control.	2
Low: Isolated failures. 1 in 4,000 is expected. Surrogate Cpk > 1.00 and is statistically in control.	3
Moderate: Occasional failures. Less than 1 in 1000/400/80 expected. Surrogate Cpk < 1.00 but in statistical control.	4 5 6
High: Frequently fails. 1 in 40/20 is expected to fail. Process is not in statistical control.	7 8

Very high: Failure is almost inevitable. 1 out of every 8/2 is expected to fail.	9 10

Column 7   List the *current controls* that will be in place to prevent or detect each failure mode. In-process inspection, variable and attribute statistical process control (SPC), fixture full-proofing, testing at subsequent operations, etc., would all be considered as *controls*.

Column 8   The next column should be labeled as *Detection*. An assessment of the likelihood that each "current control" will *detect* each failure before it leaves the production location is the purpose here. On a scale of 1–10, each "control" can be rated as follows:

Detection	*Rating*
Very high: Controls will almost certainly detect a failure. (Process automatically detects failure.)	1 2
High: Controls are likely to detect a failure.	3, 4
Moderate: Controls may detect a failure.	5, 6
Low: Controls have poor chance of detecting a failure.	7 8
Very low: Controls will probably not detect a failure.	9
Absolute certainty of nondetection: Controls will not or cannot detect a failure.	10

Reference Exhibit 9 for quick rating information.

Column 9   This column is for the calculation of a *risk priority number (RPN)*. The RPN is the product of the severity, occurrence, and detection scores. The RPNs are only evaluated against each other (i.e., a Pareto Diagram) in order to prioritize a process of continuous improvement.

Column 10   The next column is labeled as *Recommended actions*. As a result of now knowing the priorities, recommended actions can be established to reduce the severity, occurrence, and/or detection of failures.

Column 11   After the recommended action is established, list the person or area *responsible* for the action and target date.

Column 12   Here is where the actual follow-up of *actions taken* and date completed is to be recorded.

Column 13 After actions have been taken, a *new RPN* should be
calculated.

If your customer identifies critical product characteristics, another column
should be added to depict which potential failures they have deemed critical.

Once the Process FMEA is originated, it should be used in conjunction
with the Process Flow Chart, customer requirements, and blueprint to aid in
developing inspection instructions and a Control Plan. It is extremely
important to understand that the FMEA is not just an advanced quality
planning tool but a living document intended to be used as a continuous
improvement tool. It should be made available on the production floor and
utilized by Process Engineers and Manufacturing personnel in continually
tackling the highest RPNs.

2.5   Control Plans

A Process Control Plan is developed to house all identified critical, safety/
governmental, and significant product/process characteristics. Its basic
intention is to document all current control methods of the identified characteristics. Because Control Plans vary widely in format due to various
customer expectations, it is important that you follow the instructions and
include the expected content that your particular customer(s) requires. It is
very possible (especially if your company is in the automotive and/or
pharmaceutical field) that you are required to follow two (2) or more different
formats containing very different information.

Usually, it is your customer (via blueprints and engineering standards)
that identifies critical, safety/governmental, and significant characteristics. If
this is the case, hopefully your customer engages in the practice of including
your company representatives in the characteristics determination process.
In any event, your company should definitely involve in-house Engineering,
Quality, and Manufacturing personnel in a product/ process review. In that
review, your experts should determine what they feel to be critical and
significant characteristics with the aids of the blueprint, Process FMEA, and
Process Flow Chart. Refer to Exhibits 10 and 11.

Next, it is necessary to agree upon the control/measurement methods
and identify the following: inspection equipment, responsibility, statistical
techniques to be used, testing/inspection/data gathering frequencies, and
sample sizes. These characteristics can be added to the Control Plan. In the

case that your customer is not informing you of what they deem critical/ significant, these characteristics and controls alone would make up your Control Plan.

The Control Plan is a living document and should be available on the manufacturing floor when the product is being made. The Plan should be reviewed at an established frequency. Whenever there is an engineering change, process optimization, or preventative action as a result of a customer or internal concern, updates/modifications should be made to it. Because it is a living document, it should contain: the release date, last revision date, engineering change level, and approval signature.

2.6   New Job Launch

The purpose of formalizing a "new job launch" team process is to introduce front-line operational employees to new job programs. It is essential to ensure that the information they need is properly distributed in order to eliminate potential delays of normal quality/operational systems flowing smoothly.

The following is an *example* of how to go about documenting a formal procedure:

1. When Production Control at the pertaining plant receives a release for any new job or significant engineering changed product, they will notify Quality of: the product number, description, first production run date, and where the new product will be produced.
    A. "Significant engineering change" will be defined here as any change that will cause modification to the processes involved in producing/inspecting/shipping the finished product. Changes that affect document updates only will not be considered significant. This is because a Drawing and Change Control Procedure should automatically address the effect of those types of changes.
2. The Quality Manager will schedule a launch team meeting before the first production run. The Plant Manager, Receiving Inspector, all other Quality Inspectors/Auditors, SPC Coordinator, Operators from each shift, Floor Persons, Foremen from each shift, the Project and Process Engineers, and any others whom the Quality Manager feels necessary to attend should be invited.

3. The meeting notice should inform attendees to bring certain documents and/or exhibits to the meeting.

    A. The Quality Manager should be prepared with the control plan, process FMEA, process flow chart, inspection instructions, gaging instructions, accept/reject sample board, special gages, lab report, and material certification.

    B. The Project Engineer should bring the New Job Notification, blueprint, design FMEA, and sample parts.

    C. The Process Engineer or Plant Manager should bring the operator instructions, process set-up sheet, process performance record, a finished product shipping label, and packaging requirements.

    D. The SPC Coordinator should bring the process potential study and gage R&R results.

4. During the meeting, the Quality Manager should complete a checklist to signify completed review by all attendees of all of the documents, instructions, samples, gages, etc. Any concerns during the review should be documented also to allow timely corrections or improvement actions prior to the first production run.

5. After the review, all attendees should be required to sign the checklist to signify that all of the items have been reviewed with them and that any concerns they may have had were addressed and documented.

This is a simple enough process that modifying it to fit your employees' differing responsibilities and functions is every bit worth the few hours it would take to set up such a program.

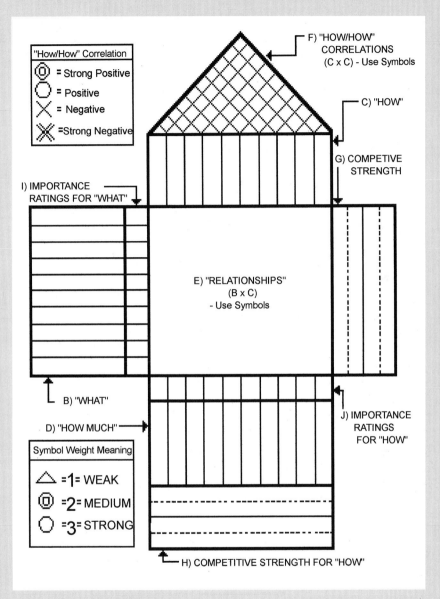

"How/How" Correlation

◎ = Strong Positive
○ = Positive
✕ = Negative
✖ =Strong Negative

F) "HOW/HOW"
CORRELATIONS
(C x C) - Use Symbols

C) "HOW"

G) COMPETIVE
STRENGTH

I) IMPORTANCE
RATINGS FOR "WHAT"

E) "RELATIONSHIPS"
(B x C)
- Use Symbols

B) "WHAT"

D) "HOW MUCH"

J) IMPORTANCE
RATINGS
FOR "HOW"

Symbol Weight Meaning

△ =**1**= WEAK
◎ =**2**= MEDIUM
○ =**3**= STRONG

H) COMPETITIVE STRENGTH FOR "HOW"

Exhibit 2: Typical house of quality

XT50-89 PROGRAM:

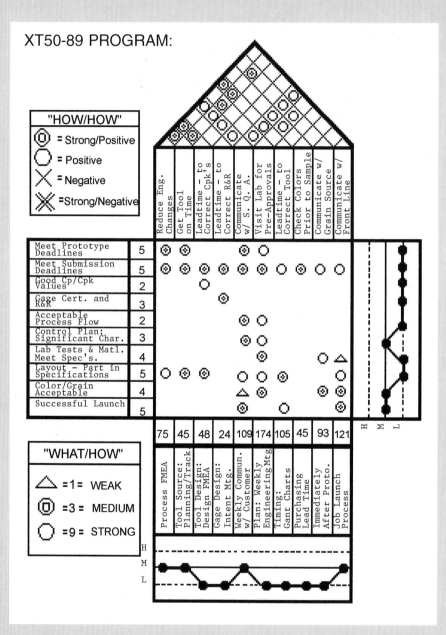

Exhibit 3: QFD house of quality

Feasibility evaluations are a series of reviews conducted by Product Engineering, Manufacturing, and Quality. The purpose is to ascertain whether a proposed design can be manufactured, assembled, tested, packaged, and shipped at acceptable levels. A feasible design must show the ability to meet production volumes and schedules, and be consistent with the ability to meet engineering requirements, quality, reliability, investment cost, unit cost, and timing objectives.

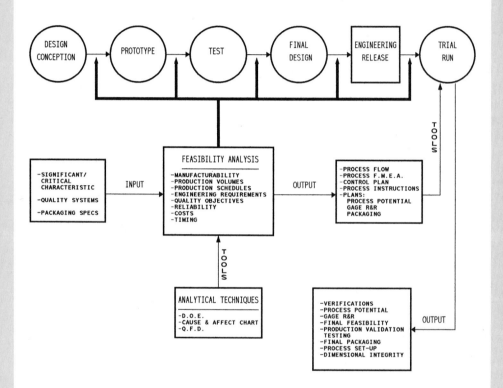

Exhibit 4: Feasibility review flow chart

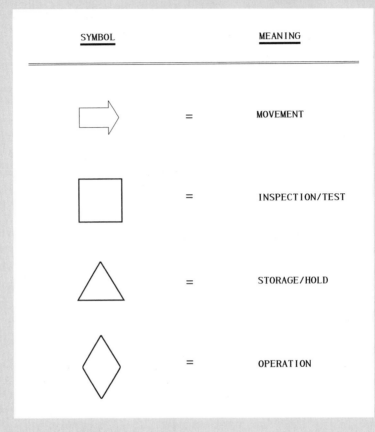

Exhibit 5: Flow chart symbols and meanings

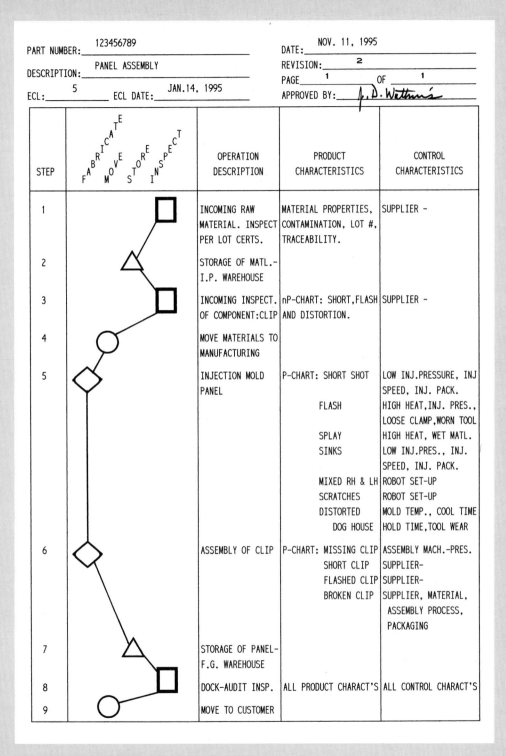

PART NUMBER:	123456789		DATE:	NOV. 11, 1995
DESCRIPTION:	PANEL ASSEMBLY		REVISION:	2
ECL: 5	ECL DATE: JAN.14, 1995		PAGE 1 OF 1
			APPROVED BY: J. D. Wethms

STEP	FABRICATE / MOVE / STORE / INSPECT	OPERATION DESCRIPTION	PRODUCT CHARACTERISTICS	CONTROL CHARACTERISTICS
1	☐ INSPECT	INCOMING RAW MATERIAL. INSPECT PER LOT CERTS.	MATERIAL PROPERTIES, CONTAMINATION, LOT #, TRACEABILITY.	SUPPLIER –
2	△ STORE	STORAGE OF MATL.- I.P. WAREHOUSE		
3	☐ INSPECT	INCOMING INSPECT. OF COMPONENT:CLIP	nP-CHART: SHORT,FLASH AND DISTORTION.	SUPPLIER –
4	○ MOVE	MOVE MATERIALS TO MANUFACTURING		
5	◇ FABRICATE	INJECTION MOLD PANEL	P-CHART: SHORT SHOT	LOW INJ.PRESSURE, INJ SPEED, INJ. PACK.
			FLASH	HIGH HEAT,INJ. PRES., LOOSE CLAMP,WORN TOOL
			SPLAY	HIGH HEAT, WET MATL.
			SINKS	LOW INJ.PRES., INJ. SPEED, INJ. PACK.
			MIXED RH & LH	ROBOT SET-UP
			SCRATCHES	ROBOT SET-UP
			DISTORTED	MOLD TEMP., COOL TIME
			DOG HOUSE	HOLD TIME,TOOL WEAR
6	◇ FABRICATE	ASSEMBLY OF CLIP	P-CHART: MISSING CLIP	ASSEMBLY MACH.-PRES.
			SHORT CLIP	SUPPLIER-
			FLASHED CLIP	SUPPLIER-
			BROKEN CLIP	SUPPLIER, MATERIAL, ASSEMBLY PROCESS, PACKAGING
7	△ STORE	STORAGE OF PANEL- F.G. WAREHOUSE		
8	☐ INSPECT	DOCK-AUDIT INSP.	ALL PRODUCT CHARACT'S	ALL CONTROL CHARACT'S
9	○ MOVE	MOVE TO CUSTOMER		

Exhibit 6: Process flow chart

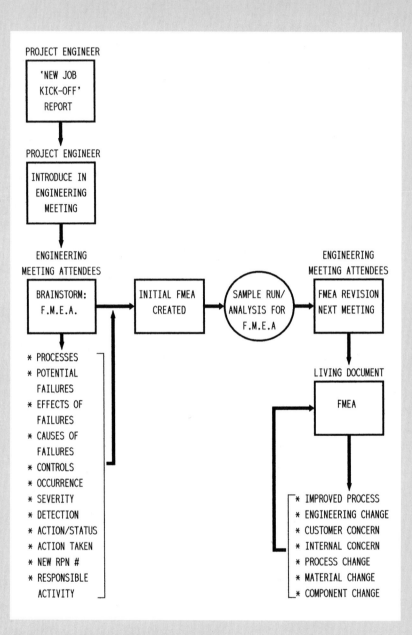

Exhibit 7: FMEA flow/procedure

COMPANY ABC, INC. SUPPLIERS MATERIAL: ABS ENGINEER JOHN D. DOE

MAIN PROCESS MOLDING SCHEDULED PRODUCTION RELEASE NOV.1995 ORIGINAL FMEA DATE JAN.1995

OTHER DEPARTMENTS NONE SCHEDULED TOOL COMPLETION FEB.1995 LAST REVISION APRIL 15, 1995

DESCRIPTION & PART NUMBER	PROCESS DESCRIPTION	POTENTIAL FAILURE MODES	SAFETY ITEM	POTENTIAL EFFECT(S) OF THE FAILURES	SEVERITY	POTENTIAL CAUSE(S) OF THE FAILURES	OCCURRENCE	CURRENT CONTROLS	DETECTION	RISK PRIORITY NUMBER (RPN)	RECOMMENDED ACTION(S) AND STATUS	ACTION(S) TAKEN	SEVERITY	OCCURRENCE	DETECTION	RISK PRIORITY NUMBER (RPN)	CAPTAIN/ DEPARTMENT
								SURROGATE CONDITIONS					RESULTING				
GADGET: P/N 1234567	MOLD GADGET	UNDERSIZE SLOT	N	TIGHT OR UNABLE TO ASSEMBLE	7	MOLD TEMP. TOO HIGH	3	X&R CHART ON SLOT SIZE-PROCESS VERIFICATION	3	54	PROVE SLOT CpK AND SWITCH TO X&R ON MOLD TEMP.AND HOLD TIME	COMPLETED 4-15-95	7	2	2	28	JANE/ MANUFACT.
						NOT ENOUGH HOLD TIME	3			54			7	2	2	28	
						TOOL WEAR	2	P-CHART	2	42							
						FLASH	3		2	42							
		OVERSIZE SLOT	N	LOOSE ASSEMBLY	5	MOLD TEMPER. TOO LOW	3	X&R CHART ON SLOT SIZE-PROCESS VERIFICATION	3	45	PROVE SLOT CpK AND SWITCH TO X&R ON MOLD TEMP.AND HOLD TIME	COMPLETED 4-15-95	5	2	2	20	JANE/ MANUFACT.
						TOO MUCH HOLD TIME	2			30			5	2	2	20	
						GOUGES DUE	4	P-CHART	2	40				4	2	40	
		FLASH	N	EDGE SHARP OVERSIZE/ UNDERSIZE SLOTS	7	HIGH HEATS	3	X&R CHART ON SLOT SIZE-PROCESS VERIFICATION	2	35	OPEN – FOLLOW PROCEDURE FOR OUT OF CONTROL CONDITIONS	OPEN					
						HIGH INJ.PRS	3	P-CHART		35							
						LOOSE CLAMP	3			35							
						WORN PT.LINE	2	PREV. MAINT & INCOMING MATL.MP CH		28							
						HIGH MATL.MP	3			35							
		SHORTS	N	POOR APPEARANCE & POSSIBLY UNUSABLE	7	LOW HEATS	3	P-CHART, PROCESS VERIFICAT. & INCOMING MATL.MP CH	2	35	OPEN FOLLOW PROCEDURE FOR OUT OF CONTROL CONDITIONS	OPEN					
						LOW INJ.PRES	3			35							
						LOW INJ.SPD	3			35							
						LOW INJ.PACK	3			35							
						LOW MATL.MP	3			35							
		SPLAY	N	POOR APPEARANCE	3	HIGH HEATS	3	P-CHART, PROCESS VERIFICAT. & MATERIAL MOIST.ANAL	2	12	OPEN – FOLLOW PROC.FOR OUT OF CONT CONDITIONS	OPEN					
						WET MATERIAL	3			18							

Exhibit 8: Process FMEA

Severity	Rating	Meaning
SEVERITY OF THE EFFECT OF THE FAILURE ON THE CUSTOMER	1	MINOR – Customer would probably not notice the effect.
	2,3	LOW – Slight inconvenience to customer.
	4,5,6	MODERATE – Failure would inconvenience or annoy customer.
	7,8	HIGH – Not a safety issue but would cause serious disruption (inoperable).
	9,10	VERY HIGH – Failure affects safety of end-user or involves government regulation noncompliance.

Occurrence	Rating	Meaning
OCCURENCE OF FAILURE	1	REMOTE – Failure not likely. Less than 1 in a million expected.
	2	VERY LOW – Isolated failures. 1 out of every 20,000 expected.
	3	LOW – Isolated failures. 1 out of every 4,000 expected.
	4,5,6	MODERATE – Occasional failures. 1 out of every 1000/400/80 (respectively) expected.
	7,8	HIGH – Frequent failures. 1 out of every 40/20 (respectively) expected.
	9,10	VERY HIGH – Failure almost inevitable. 1 out of every 8/2 (respectively) expected.

Detection	Rating	Meaning
CURRENT CONTROLS THAT WOULD ENABLE DETECTION OF THE FAILURE	1,2	VERY HIGH – Controls will almost certainly detect a failure (process will automatically detect a failure).
	3,4	HIGH – Controls are likely to detect a failure. (100% INSPECTION)
	5,6	MODERATE – Controls may detect a failure. (X&R CHARTING)
	7,8	LOW – Poor chance of detecting a failure.
	9	VERY LOW – Probably will not detect a failure.
	10	ABSOLUTE CERTAINTY OF NON-DETECTION – Will not or cannot detect a failure.

Exhibit 9: Process FMEA ratings summary

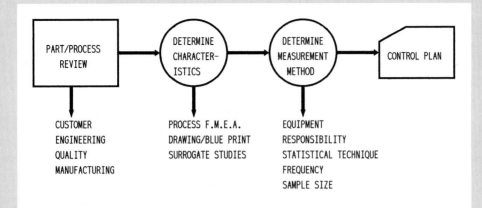

Exhibit 10: Significant/critical characteristics determination

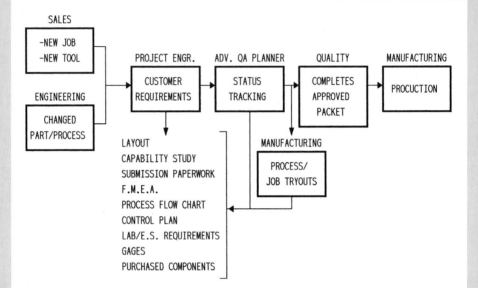

SALES
```
┌──────────────┐
│ -NEW JOB     │
│ -NEW TOOL    │
└──────────────┘
```

ENGINEERING
```
┌──────────────┐
│ CHANGED      │
│ PART/PROCESS │
└──────────────┘
```

PROJECT ENGR.
```
┌──────────────┐
│ CUSTOMER     │
│ REQUIREMENTS │
└──────────────┘
```

ADV. QA PLANNER
```
┌──────────────┐
│ STATUS       │
│ TRACKING     │
└──────────────┘
```

QUALITY
```
┌──────────────┐
│ COMPLETES    │
│ APPROVED     │
│ PACKET       │
└──────────────┘
```

MANUFACTURING
```
┌──────────────┐
│ PROCURTION   │
└──────────────┘
```

LAYOUT
CAPABILITY STUDY
SUBMISSION PAPERWORK
F.M.E.A.
PROCESS FLOW CHART
CONTROL PLAN
LAB/E.S. REQUIREMENTS
GAGES
PURCHASED COMPONENTS

MANUFACTURING
```
┌──────────────┐
│ PROCESS/     │
│ JOB TRYOUTS  │
└──────────────┘
```

RESPONSIBILITIES

REQUIREMENT	ENGINEERING	QUALITY ASSURANCE	MANUFACTURING
LAYOUT	--------------	LAB TECHNICIAN	----------------
CAPABILITY	--------------	S.P.C. TECHNICIAN	----------------
SUBMISSION	ADV. QA PLANNER	Q.A. MANAGER (APPROVE)	----------------
F.M.E.A.	PROJECT ENGR.	SPC TECH./Q.A. MANAGER	FOREMAN/MANAGER
FLOW CHART	PROJECT ENGR.	SPC TECH./Q.A. MANAGER	FOREMAN/MANAGER
CONTROL PLAN	PROJECT ENGR.	LAYOUT/Q.A. MANAGER	----------------
LAB/E.S. REQ.	--------------	LAB TECHNICIAN	----------------
GAGES/FIXTURES	PROJECT ENGR.	LAB TECHNICIAN	----------------
PURCH.COMPON.	PROJECT ENGR.	LAB TECHNICIAN	PURCHASING AGENT
TRYOUTS	PROJECT ENGR.	INSPECTOR/Q.A. MANAGER	PRODUCTION CONTROL

Exhibit 11: Quality planning (requirements and responsibilities)

3   Sampling and Submissions

3.1   Prototype Sampling and Submission

The purpose of fabricating a prototype product is to test and evaluate the *design* for function ability and to ensure that all engineering requirements can be met. This procedure also aids in fine-tuning the production tool design and process intent. Prototypes should be made with the raw material that the blueprint specifies and that will be intended for production.

The Quality function should, (at a minimum), inspect all dimensional characteristics as shown on the product's blueprint and test the product to all required material specifications. If your customers have prototype submission guidelines, then of course, your company would be required to adhere to those. Some customers, for example, require that every single prototype sample shipped to them be 100% dimensionally laid-out and/or tested. Also, most customers require that a warrant (typically of their format) accompany all shipped prototype samples.

No prototype should ever be shipped to a customer without both Engineering's and Quality's approval. Internal procedures should be documented and included in the Quality Manual that detail specific responsibilities and the process of Quality's involvement in regard to the inspection, testing, approval, and shipment of all prototype material.

3.2   Production Sample Run Tryouts

A manufacturing sample run should be scheduled immediately after the completion of the production tool/process. I recommend the following procedure to assure meaningful sample runs:

3.2.1   Sample Quantities

The Project Engineer should issue a sample tryout notification to all departments. On the notification form, it should be specified that at least 300 samples be produced during the run and that the last 50 consecutive samples be serialized with a grease pencil or marker. With this information, Production Control can schedule the run.

3.2.2   Deviation from Sample Quantities

In the event of processing, quality, or tooling problems occurring during the run, the Project Engineer (who should be present) can revise the quantity of parts needed. If for some reason the Project Engineer cannot be present and problems arise that would affect the dimensional integrity or aesthetic expectations of the product, at least 20 samples should be produced, if possible, and the process halted. Those 20 samples should be labeled "ATTENTION ENGINEERING" and forwarded immediately to a designated sample storage area.

3.2.3   Sample Identification and Distribution

After 300 samples are produced, the 50 consecutively marked samples should be packaged separately and labeled "ATTENTION SPC". The remaining 250 samples should be labeled "ATTENTION ENGINEERING AND QUALITY." There should be a form label specifically for samples with the word "SAMPLE" at the top. It should provide spaces for manufacturing to enter a product number, engineering change level, and description (at minimum) in addition to an "ATTENTION" space for proper distribution.

3.2.4   Sample Storage Area

Once labelled, all samples should be taken immediately to a designated sample storage area. This is very important because if samples are precariously taken to Engineering, Quality, or a warehouse, it is inevitable at some point that they will disappear or wind up at your customer's location before approval. The sample storage area should be located away from the normal flow of production and finished goods. It is wise to have the area enclosed and only accessible to those who need entry.

3.2.5   Sample Storage Records and Activity

To know current inventories, for traceability, and to aid in the moving or disposal of samples, a log sheet should be maintained at the storage area. The log should contain such information as:

1. Product number,
2. engineering change level,
3. quantity of samples entered and date,
4. initials of the person who brought the samples to the area,
5. quantity borrowed and date,
6. initials of the borrower,
7. balance, and
8. disposition, initials, and date.

Areas 1, 2, 3, and 4 on the log sheet should be filled out by the person responsible for delivering samples to the storage area. This will allow any individual to know the contents of the area without having to enter the area, move containers, etc. Areas 5 and 6 on the log should be in triplicate to allow more than one entry here for each product number. These spaces should be filled out by the Engineer (evaluation), Lab/Reliability (testing), SPC Coordinator (Process Potential Study—50 consecutively marked samples), Metrology Technician (dimensional layout), etc., who may need to remove samples for any reason. If samples are returned after they are borrowed, a line can be drawn through the borrower's entry on the log. Areas 7 and 8 should be filled out by the engineer or other responsible individual when the samples can be moved to an in-process or finished goods warehouse or dispositioned otherwise.

3.2.6   Maintenance of Storage Area

In regard to areas 7 and 8 on the sample storage area's log sheet, to depend on a variety of people to see to the removal of materials from the area will probably result in some chaos and a very messy area. An effective approach to consistency here may involve a more in-depth process such as the following: Put a Quality Department individual in charge of the area who will make copies of the log sheet(s) at the same frequency of your Engineering meeting. The copies can be distributed to the engineers at the end of the meeting, and the appointed Quality individual will go through the list and ask for recommended dispositions of samples that are not at recent engineering change levels, have already been submitted and approved, etc. The engineers will instruct him/her to either dispose, recycle, move, ship, rework, etc., the samples, and the Quality Department member can make the entries on the master log sheet that are applicable. At the end of the meeting, a copy of the master log can be made and those samples needing to be removed from storage can be highlighted and forwarded to the manufacturing supervisor/manager or whomever is to be responsible for carrying out the disposition.

3.3   Sample Submissions to the Customer or Production Approval Source

All sample submissions for production approval should be submitted in accordance with the requirements set forth by the customer or approval source. Typically, the following items must be completed in order to obtain production approval from the customer or production approval source:

- Minimum 300 cycles sample run,
- laboratory report that breaks down all required specifications/ testing requirements and shows that all criteria was met,
- Process Flow Chart,
- Process Control Plan,
- Process Potential Studies showing that the $Cp > 1.67$,
- Gage R&R Studies showing less than 10–20% error,
- Process FMEA,
- material certifications,
- Gage certifications, and
- dimensional layout showing all blueprint dimensions to be within specification tolerances.

Usually, in addition to these documents, you will also submit samples, blueprints, auxiliary drawings, and gage/holding fixtures for approval.

Sample submissions should be the responsibility of the Quality Department. It is important that one individual or group champion the task of compiling what is necessary to submit packages that will be approved. Refer to Exhibit 12 to view a typical submission process flow and Exhibit 13 for an example of a packet cover sheet.

3.3.1   Records

A "master" of the product that is representative of the samples submitted should be tagged as such and retained for the life of the product and five years thereafter if it is a safety-related product. If you have a customer that splits up the submission process into two or more phases (i.e., ungrained part submission and then a grained part submission after dimensional approval, or a "natural" material submission and colored material thereafter), representative samples from each phase should be labeled and retained as "masters". The retained masters will come in handy if a dimension, attribute, or performance quality is found later to be out of specification. The label on the master should contain the submission date, engineering change level, production approval date, and any other traceability information.

All documents that are submitted should first be duplicated and arranged exactly as the originals and bound together. The duplicate submission packet should be filed and retained for the life of the product plus two years (five years for safety-related products).

3.3.2   Submission Tracking

There should be a checklist made up of all submission requirements to be used in the event of any sample submission, (refer to Exhibit 14). Someone should be assigned responsibility of maintaining the checklists and assuring that all requirements are progressing as needed to meet deadlines. The checklist should be headed with spaces for the product number, description, customer, engineering change level and date, the type of submission (i.e., new job, change, new color), the due date, and the date that the tracking report was initiated. A list of all requirements should follow with the expected due date, date completed, and comments for each individual requirement. This form can serve as a progress summary in engineering meetings, meeting deadlines, as well as a problem assessment tool to determine where hold-ups are occurring so that preventative actions can be taken to continuously improve the submission process.

As part of an improvement tracking process, graphs could be reviewed monthly that depict the percentage of "on-time submissions" and "first-time approved submissions."

3.4   Pilot Run and Submission

Manufacturing product for a *pilot* phase or shipment should be conducted exactly as it would be in normal production. The product should not have to be reworked and all documentation should be finalized and available at the place of manufacture (i.e., operator and inspection instructions, process FMEA and Flow Chart, etc.).

Your company should have a written procedure that defines who is responsible for reporting (with specified timing) that a scheduled run will produce pilot parts. The procedure should detail all requirements of your company and its customers in regard to pilot runs and shipments (i.e., warrants, special labels, deviations, etc.).

I suggest that all pilot runs be 100% inspected by Quality and approved by the Quality Manager, or another individual who supervises the inspector, who will approve the product once in production. That person should sign or initial the package or shipping label to indicate product pilot approval.

In the event that the product deviates from the blueprint, engineering specifications/standards, or has not been production-approved, the procedure should define the process to be followed and the responsible department/employee.

If pilot product can be taken from the original sample run of 300 pieces or more, special attention should be paid to the engineering release level. In the event that engineering changes have occurred and older level samples were not removed from the system for some reason, a serious mistake could be made. It is for this reason that *all* previous level samples be identified as "reject" (reason: obsolete) if they do need to be saved.

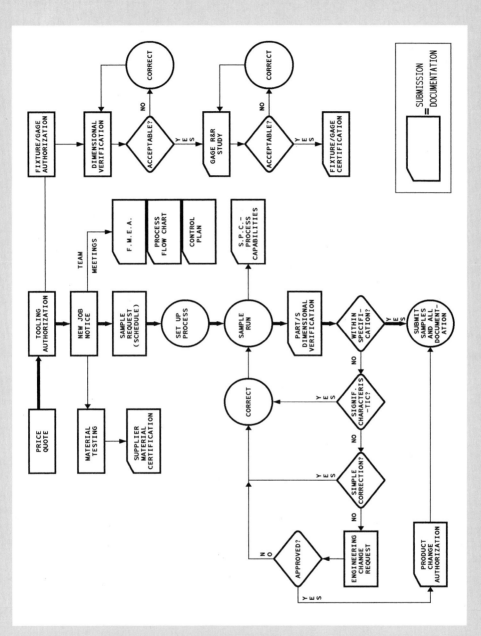

Exhibit 12: Typical submission process

SEND TO:

_____

_____

_____

_____

ATTENTION:_____

PART NUMBER _____

DESCRIPTION _____

BLUE PRINT NUMBER _____

ECL _____ DATE_____

RESULTS:

The inspection results for all dimensions, other specifications, and laboratory tests such as chemical, physical, metallurigical, etc., show that the samples meet all requirements... ☐YES ☐NO If 'NO', the variations are to be noted on checked print/s, inspection results sheets, and laboratory records.

SUPPLIER NAME   _____

SUPPLIER ADDRESS _____

PRODUCING LOCATION (IF DIFFERENT) _____

SUPPLIER CONTACT/TITLE _____ PHONE _____

REASON FOR   ☐New Part   ☐Process Change   ☐Tool Transfer   ☐New/Additional Tool
SUBMISSION: ☐Engineering Change   ☐Correction to Discrepancy   ☐Material Change to Optional
               ☐Replaces Part Number: _____
               ☐Other: _____

SUBMISSION   ☐REQUIRED NUMBER OF SAMPLES   ☐INSPECTION RESULTS   ☐CHECKED BLUEPRINT
CHECKLIST: ☐LAB AND TEST RESULTS   ☐S.P.C./CAPABILITY STUDIES   ☐PROCESS CONTROL PLAN
               ☐GAGE R&R REPORT   ☐PROCESS FLOW CHART   ☐PROCESS F.M.E.A.   ☐CERTIFICATIONS

DECLARATION:

I affirm that the samples represented by this warrant are representative of the parts used to determine all attached results. The samples were manufactured according to the process flow chart and meet all requirements as specified on the blueprint.

_____   _____

AUTHORIZED SIGNATURE/TITLE   DATE

              DO NOT WRITE BELOW THIS LINE – CUSTOMER OR PRODUCTION APPROVAL SOURCE USE ONLY

☐FULL PRODUCTION APPROVAL
☐LIMITED PRODUCTION APPROVAL   _____   _____
☐REJECTED FOR PRODUCTION   AUTHORIZED SIGNATURE/TITLE   DATE

Exhibit 13: Supplier submission for production approval (cover sheet)

THIS REPORT IS TO BE INITIATED IN THE EVENT THAT A SAMPLE SUBMISSION IS TO BE MADE FOR APPROVAL TO PROCEED WITH PRODUCTION. QUALITY ASSURANCE INITIATES THIS REPORT UPON RECEIPT OF NEW JOB NOTIFICATION OR ECN FROM ENGINEERING AND WILL REVIEW IT WEEKLY IN THE ENGINEERING MEETING.

PART NUMBER:_____   DESCRIPTION:_____

ECL:_____ ECL DATE:_____   CUSTOMER:_____   DATE INITIATED:_____

TYPE OF SUBMISSION:_____   SUBMISSION DUE DATE:_____

REQUIREMENTS	DATE EXPECTED	DATE COMPLETED	COMMENTS
BLUEPRINTS RECEIVED BY METROLOGY	_____	_____	_____
55 SAMPLE PARTS FROM A 300 CYCLE RUN	_____	_____	_____
PROCESS FLOW CHART	_____	_____	_____
PROCESS F.M.E.A.	_____	_____	_____
PROCESS CONTROL PLAN	_____	_____	_____
GAGE CERTIFICATION	_____	_____	_____
GAGE R & R	_____	_____	_____
S.P.C.- PROCESS POTENTIAL STUDY	_____	_____	_____
LABORATORY REPORT	_____	_____	_____
E.S. TESTING	_____	_____	_____
MATERIAL CERTIFICATION/S	_____	_____	_____
DIMENSIONAL LAYOUT	_____	_____	_____
BALLOONED/MARKED PRINTS	_____	_____	_____
PURCHASED COMPONENTS	_____	_____	_____
SAMPLES RECEIVED	_____	_____	_____
DIMENSIONAL LAYOUT	_____	_____	_____
LAB REPORT/CERTIFICATION	_____	_____	_____
CONTROL PLAN	_____	_____	_____
PROCESS F.M.E.A.	_____	_____	_____
GAGE CERTIFICATION	_____	_____	_____
GAGE R&R	_____	_____	_____

ADDITIONAL COMMENTS:_____

NOTE: IF ONE OF THE ABOVE ITEMS IN NOT REQUIRED, 'NR' IS TO BE PLACED IN BOTH THE 'EXPECTED' AND 'COMPLETED' COLUMNS. UPON THE SUBMISSION DUE DATE, IF ANY 'DATE COMPLETED' LINE IS BLANK, THE SUBMISSION WILL NOT BE APPROVED BY QUALITY ASSURANCE TO BE SUBMITTED TO THE PRODUCTION APPROVAL SOURCE. ACTIONS MUST BE TAKEN IN THE EVENT OF MISSED DATES AND REPORTED TO THE CUSTOMER PURCHASING ACTIVITY IMMEDIATELY.

Exhibit 14: Sample submission status report/checklist

4   Supplier Development and Control

4.1   Supplier Guidelines

A supplier or subsource may not meet your expectations if they do not know what your expectations are. The time and effort involved in putting together a "Supplier Guidelines" handbook or manual is priceless. The guidelines manual should contain your company's expectations in all areas. Some of those areas covered may be as follows:

- Advanced quality planning,
- sample submission,
- statistical process control,
- acceptance standards,
- certification/warranty,
- record retention,
- education and training,
- nonconforming product,
- supplier performance tracking,
- annual product validation,
- eliminating incoming inspection,
- supplier surveys,
- on-time delivery,
- routing instructions,
- advanced shipping notification,
- CUM reconciliation,
- rejections—CUM reductions,
- service,
- cost savings, and
- packaging/labeling.

The manual or handbook might also include such exhibits as:

- Supplier warrant of material for production approval,
- supplier performance report form,
- blanket order release example,
- SPC report form,
- supplier routing instruction,
- reject notice or form,
- ASN example,
- corrective action form,
- CUM rollover example,
- on-sight survey form, and
- contact list.

The guidelines could be documented in list format, focusing on the areas where, in the past, there have been communication problems with the supplier base. There need not be a full blown procedure printed for each requirement your company expects the supplier to meet. I also suggest that Purchasing and Quality (at a minimum) get together and write a mission statement for "Supplier Relations." That mission statement could then be the very first page of the guideline manual with the Table of Contents following.

Once a guidelines book is put together, the Purchasing Manager and Quality Manager should set up an orientation for all suppliers in order to introduce them to the new guidelines book and to review the details of the requirements listed therein. After that, whenever there is a significant increase of suppliers to the base or changes to the guidelines set forth, orientations should be held for the suppliers affected. It is a sensible way to answer questions of the suppliers and let them tour your facility. I have found that even if a supplier is much bigger than their customer in terms of reputation, size, and profit, they still appreciate having requirements communicated to them in this fashion. Even material distributors can learn what you expect of their suppliers.

Purchasing should have the manuals or handbooks always available to give to new suppliers and potential suppliers. As well, the manual should be maintained, updated, and improved by Purchasing and Quality.

4.2   Supplier Assessments

Assessments or surveys should be conducted on your suppliers for three main reasons:

1. To be assured that the supplier can meet your company's requirements,
2. to learn the direction your company should go to develop that supplier, and
3. to serve as a guideline in selecting new suppliers.

Because of reason number 2, it is important that assessments occur at regular intervals.

A simple way to establish the criteria to be met by a supplier in an assessment is to base it very similarly to your customers' assessment criteria to which they expect your company to comply. If your customers do not conduct supplier assessments, then base the criteria on what you expect of your own company. The reason that I first suggest mirroring a customer assessment is that it could also serve as an audit for your company to conduct on itself.

Once the criteria are established (i.e., Quality, Cost, Delivery, Technology/Engineering, Leadership/Management, etc. requirements), a numerical rating system should be applied. For example: 5 points for each requirement that a supplier exceeds, 4 points for meeting the requirement, 3 points if minor improvements are needed, 2 points if major improvements are needed, and 1 point if the requirement in no way is met. An overall score for each section could then be tallied for quick reference as to the supplier's abilities in particular areas.

You may opt to set up various categories for suppliers so that on-sight assessments can be conducted on those suppliers that need one most. Some examples of categories are as follows:

1. *Standing score:*   This category could be for those suppliers who were assessed by your company within the last two years and
    a. had received an acceptable assessment score, and/or
    b. had acceptable performance scores (refer to Section 4.5, Supplier Performance Tracking and Reporting).
2. *Acceptable by customer standards*:   This category could house the names of all of your suppliers who have had an assessment conducted at their facility by one of your company's customers. It

would be redundant for your company to spend time and effort to assess a supplier that one of your customers have already assessed and found them adequately meeting requirements. I suggest that you request actual copies of those assessments from your suppliers for your records. Do not place suppliers in this category if the assessment score that your customer gave them was unacceptable. Obviously, if this were the situation, that supplier should be able to use any direction and advice that you could offer during an additional assessment.

3. *Self-assessment.* If you have suppliers who have exemplary supplier performance scores in the areas of Quality, Service, and Cost, you can justify placing them in a "self-assessment" category. Placement of a supplier in this category would result in that supplier receiving a copy of your company's assessment form and procedure for them to fill out and return.

4. *On-sight.* In this category, place all other suppliers. Those suppliers will require that an assigned assessor or assessment team from your company conduct the audit at their facility.

It is a good idea to draw up a Gantt chart at the beginning of each year to list all of the suppliers to be assessed at their plant (refer to Exhibit 15). Include who will be responsible for assessing them and when they will be assessed. On this same Gantt chart, at the bottom, list the other categories and the suppliers who fall under each one. A chart such as this will make a great quick reference sheet for purchasing as well as the individuals in your company who may be scheduled to conduct assessments. The Gantt chart should be reviewed monthly by Purchasing, Quality, and any other affected department to track audit results and assure that they are occurring when they are scheduled.

Improvement Action Plans should be requested of all "on-sight" and "self-assessment" suppliers who score less than perfect ratings. The supplier should be instructed to reference each requirement that received less than a perfect score, the assessor's comment, the planned improvement action, who's responsible, and the target date when documenting their plan. Target dates, in some cases, may need to be mandated by the assessor. The time period in which the plan should be written and in the hands of the assessor should also be relayed to the supplier. The assessor should be held responsible for following up on corrective actions, as well as forwarding the originals of the actual assessment, comments, and action plans to Purchasing.

4.3   Receiving Inspection

Receiving inspection is slowly becoming an activity of the past. Through such programs as "Reduced In-House Inspection" and "Supplier Warrants and SPC" (in this chapter), many companies have significantly reduced or eliminated the actual inspection of incoming materials, components, etc. However, it is difficult to put such programs into place if your company does not have (or needs major improvement in) a current receiving inspection program.

4.3.1   Designated Receiving Hold Area

Basically, all received goods that will become part of the products your company produces should be taken directly to a staging area where they are segregated from other stock until approved by Quality. That area should be marked and labeled as the "Receiving Hold Area" or similarly and maintained by the Receiving Department.

4.3.2   Inspection

All of those purchased materials should be subject to some type of inspection by Quality. In many companies, there is a "Receiving Inspector" in the Quality Department. The inspector would verify conformance to the requirements of the purchase order, contract, specification, drawing, waivers, etc. The purchase order should require that material certifications accompany or be received before approval of all purchased lots of material.

4.3.3   Inspection Instructions

The inspector should follow written inspection instructions for each purchased material. Those instructions should include (at a minimum) the following:

- Supplier name,
- material description,
- color (if applicable),

- material specification (from blueprint),
- material number or nomenclature,
- customer(s),
- acceptance standard criteria,
- accept/reject instructions,
- release date,
- last revision date, and
- approval signature.

Acceptance Standard Criteria should always consist of the mandatory recording of the material lot number, quantity received, quantity inspected, and the receipt of a material certification. In addition to those, you would also list all other inspections and/or tests to be completed (i.e., impact test, color match, melt flow rate, filler rate, length, diameter, etc.).

A sampling plan in accordance with zero discrepancies should be specified on the inspection instructions or at least referred to.

4.3.4   Inspection Records

Detailed results of the inspection should be recorded on a log sheet. The log sheet procedure should not allow the use of check marks or "OKs" to denote the approval of each inspection or test. The log should show exactly how many or how much was inspected, and exactly how many or how much was accepted or rejected for each inspection or test. In other words, if recording the results of inspection for standards A, B, and C, it might be recorded that 5 samples were inspected and the following amount approved: A–5, B–5, C–4. The log sheet would indicate that the 5 samples checked passed both A and B standards, but 1 sample failed standard C.

I have found that of all types of inspection records, the Receiving records would be best kept on computer for quick traceability of lot numbers. If you have bar-coding throughout your process, you would not need a "dead" document on the computer and instead would use a log that could be filed.

4.3.5   Approval/Rejection

All incoming goods that are deemed acceptable should be marked as so. This could involve the application of a stamp, label, etc. to the container. After the goods are marked as being "approved," they should be moved to the in-process warehouse or staging area.

If a rejection of a lot results from incoming inspection, information of the rejection should be documented on a form that requires (at least) the following: Product number and description, supplier, date received, amount received, amount inspected, amount discrepant, amount rejected, discrepancy details, inspector, date of report, and disposition. Quality should then make sure that the goods are clearly marked as being "rejected" by the use of the application of a label, stamp, etc. on each container. Copies of the rejection reports should be immediately forwarded to Purchasing and Production Control. Purchasing is typically responsible for forwarding the rejection report to the supplier and requesting immediate interim and corrective actions.

When a rejection of a lot based on a certain percent of product found to be discrepant would cause curtailment of production, a 100% inspection may need to be performed until a sufficient amount of acceptable product has been segregated to assure continued production. Usually when this occurs, the company would charge the supplier for down-time and sort-time.

If purchased goods that are approved by Quality and released to production are later found to be discrepant, the same rejection procedures should occur. The receiving inspection log sheet should be updated to reflect the latter discrepancy and indicate it as having been detected in-process.

4.4   Supplier Statistical Process Control (SPC)

SPC data can be gathered by the supplier and forwarded to your company for evaluation. The gathering of data for SPC charts in Receiving Inspection can be somewhat erratic due to the fact that there are no set frequencies for inspection. As well you will not necessarily receive sequential shipments of lot numbers from suppliers as they may ship the same product to other customers.

For every material and in-process product received, there should be a copy of the supplier's control plan on file. The control plan will indicate what is being inspected at the supplier location, and will show if characteristics, properties, etc., are charted with SPC. If the supplier is advanced in SPC and performing it on their process rather than the product, they should be able to justify how and why particular process parameters were selected over others to be monitored.

4.4.1   Procedures

Before you begin a supplier SPC program, a formal procedure should be written. The procedure should address the method and system by which all incoming supplier SPC would be received, evaluated, and reported. It should specify who is responsible for each phase of the process and the frequencies that each phase should occur.

4.4.2   Supplier Submission of SPC Data

A letter can be written from the Quality Manager, SPC Coordinator, or Purchasing Manager to the supplier Quality Manager and copied to their Sales Manager to introduce your requirement of SPC data. The data can be requested at a frequency that is comfortable for both your company and your supplier. I believe that a request for data every shipment is completely unnecessary and will cause a great deal of frustration to those who have to organize the program. Every month or every quarter is usually a good frequency to begin with.

   To aid both parties in the organization of the information, I suggest that your company develop a form onto which the supplier can group the important SPC results and submit with the actual data. The form could be sent along with the initial letter and each reminder letter when due. Aside from the obvious (i.e., supplier name, product number, etc.), the form should ask for the report month or quarter, and the Control Plan date (so that you can be assured you have the latest on file). The body of the form can house rows and columns headed as:

- Characteristic,
- specification,
- tolerance,
- chart type (X&R, nP, P, c, u, etc.),
- subgroup size,
- number of subgroups,
- in control (yes/no block),
- average,
- range,
- Cp value, and
- Cpk value.

At the bottom of the form can be placed more yes/no blocks to aid the supplier in following your procedure.

Example:

Y/N _____   If any characteristic of this product has a Cpk less than 1.33, are you 100% sorting and attaching a copy of your corrective action plan?

Y/N _____   If any characteristic has a Cpk above 1.33, is there a continuous improvement plan in place?

If you have problems with suppliers complying to your requirement of SPC data, call them and set up an appointment for a meeting so that you can begin developing them in using the tool. If it is simply an issue of them not wanting to comply, have your Purchasing Manager handle it through Sales. You could make sure the request for SPC is on the purchase order and return goods to the supplier if the data is not received or is not on time.

4.4.3   Evaluating and Reporting on Received SPC Data

If, for instance, your company was receiving SPC every quarter, your procedure may detail the following points:

- _____ is responsible for maintaining a tracking log that lists each supplier and the information reported on the SPC form.
- _____ will compare the back-up data, charts, and histogram to the submitted SPC form and verify accuracy.
- _____ will then create a report to be copied on a quarterly basis to the Quality Manager, Purchasing Manager, and Improvement Tracking Team. The report will categorize

  1. Suppliers whose process shows statistical control and acceptable product capability,
  2. suppliers whose product(s) are within specification but process is out of control,
  3. suppliers whose process is in control but product(s) out of specification,
  4. suppliers whose process is out of control and not capable of producing product within specification, and

5. suppliers who did not respond by submitting current quarter SPC forms and data.

• _____ will be responsible for following up on the action plans and progress of those suppliers in category 2, 3, and 4 on the quarterly Supplier SPC Report, and the Purchasing Manager will contact suppliers who fall into category 5.

• _____ will report the status of this program quarterly in the regularly scheduled SPC meeting.

Once your program is off the ground, you may decide to eliminate the need for suppliers to send all of the raw data, control charts and histograms along with the SPC form. When your "evaluator" has established a good confidence level in suppliers as to the validity and accuracy of their reporting, he/she could justify requiring only the SPC form quarterly.

4.5   Supplier Performance Tracking and Reporting

Suppliers' performance can be tracked in such areas as quality, service, and cost. Well documented tracking records can be used to guide your purchasing and quality departments in making sourcing decisions and in identifying priorities in supplier development efforts.

4.5.1   How to Begin a Tracking Program

The first step is to get all of the necessary disciplines together (Purchasing, Quality, Manufacturing, Engineering, and Production Control) to discuss the idea. Once the team has listed those areas that are important to them in regard to the supply base, they should define what in each area is major and apply weights or ratings to each identifying its significance. It should also be decided at what frequency the tracked data would be reported (i.e., monthly, quarterly, etc.).

After the categories, items, and importance ratings have been agreed upon, the team should begin to develop the tracking process. Once each item has a method for the data to be collected and an individual assigned responsibility for it, a report log should be constructed that will house all of the information (please reference Exhibit 16). The report log could be put on a computer network to save paper shuffling and time if your company has such a system. If not, the report log could be circulated at report time to each individual responsible for contributing data. When the log reaches each person, they will enter the information in a provided space labeled with the area that they are responsible for tracking and pass it on to the next person in the loop. After the report log is complete, it can be circulated to the individual who will be responsible for applying the weighted ratings to each result and tallying an overall score.

4.5.2   Supplier Performance Reporting

Once each report is completed, there should be a score for each area, as well as an overall score. Copies of the report should be sent to each supplier. Also, depending on the frequency established for the report periods, supplier performance reviews can be held either in its own meeting or an already established meeting where all disciplines are in attendance. I suggest that a graph be constructed and maintained for each supplier that depicts the overall score for each period during the current report year. This graph will aid Purchasing and Quality when conducting assessments and will provide a quick reference report for the entire year. The supplier's last assessment scores could be included on the graph as well. Please reference Exhibit 17 for a graph example.

4.5.3   Corrective Action Process

The team, while deciding on importance ratings (weights), should determine what individual and overall scores would be considered acceptable and not acceptable. It is a good idea to document those determinations on the report that the supplier would be copied on. It should also state that a written corrective action plan be submitted to Purchasing by the supplier for any area receiving an unacceptable score or an overall unacceptable score. When Purchasing receives an action plan, he/she should copy it to the person responsible for the area that was unacceptable and to the individual who maintains the reports so that it can be noted that indeed the required action plan was received.

4.5.4   Supplier Development

In the event that a supplier is repeatedly performing below acceptable standards or their action plans appear to be ineffective, the Purchasing Manager should contact the supplier to set up a meeting at either location. Purchasing should notify Quality, Production Control, Receiving, etc., if the area(s) where the supplier is found to be performing below standard affects their area of responsibility. The purpose of the meeting will be to assist the supplier in developing a successful plan of action and to provide them with the knowledge, resources, information, training, etc., that will help them accomplish the desired results. Some examples of development are:

- Invite the supplier to a particular in-house training seminar that pertains to their current needs.
- If the supplier had an ineffective SPC program, you might set up a day for their SPC person or Quality Manager to spend a couple of days with your SPC Coordinator.
- Assist them in an action by informing them of how your company would rectify that type of concern.
- Suggest pertinent courses, seminars, books, etc., that would provide them the knowledge/training needed to accomplish corrective actions.

4.5.5   Procedures

It is important that once the Supplier Performance Tracking Program is decided upon, it be documented as a procedure. The procedure should detail all steps of tracking, reporting, acting, and responsibilities. It should also refer to the documents used in conjunction with the tracking and reporting.

4.6   Elimination of Receiving Inspection

A goal for Receiving Inspection would be to eliminate that function through supplier development. It is not cost efficient to inspect goods that have already been subjected to in-process and final approval processes. The key to meeting the goal of eliminating Receiving Inspection is to *justify* it. It is justified by developing your supplier(s) so that they understand your requirements and put the necessary systems/processes in place that will assure continuous improvement in quality. It is further justified by collecting data that proves over time that the supplier can meet your company's expectations.

4.6.1   Formal Plan

As with any long-term goal, to execute it requires a well-documented plan that details milestones, expected progress, and responsibility. Anyone who has made it this far into this book knows that "the plan" should be a product of all the great minds of a multi-disciplined team.

To begin, the team should list what is currently inspected through Receiving Inspection (i.e., SPC, visual and dimensional standards, properties, verification of warranty information, etc.). Those can be what the plan will focus on collecting and comparing data on. To eliminate warranty verification for instance, the plan could state that it is the *responsibility* of the Lab Analyst to maintain a correlation graph that depicts both the properties readings of the supplier and your company. He/she will establish the criteria for good, acceptable, fair, and negative correlations. *Quarterly*, the analyst can evaluate the data and place suppliers in one of the four categories. A supplier who remains in the good" category for three consecutive quarters may have their warranty information verified only once per year. A supplier who is in the "acceptable" category for two consecutive quarters can go on a reduced inspection program from warranty verification every shipment to every other shipment. A supplier that is placed in the "fair" category can be subject to current inspection—unchanged.

And suppliers that fall into the "negative" category should be called by Purchasing to work on an improvement action plan. The *goal* could be to reduce Receiving Inspection for warranty verification by 30% the first year, and 10% each year thereafter until it is eliminated to an annual validation. A graph can be maintained by the Lab Analyst to show the overall percent reduction of warranty verification each quarter and report the status in staff meetings.

The plan for each area could be detailed like the above example. I suggest that any plan such as this be introduced to your suppliers via a letter from the Purchasing Manager so that they understand your company's goals that they influence.

In conjunction with any plan to eliminate Receiving Inspection in your company, the Supplier Performance Tracking and Supplier Assessments Scores should be of importance and consideration in whether to reduce or eliminate any inspection.

ON—SIGHT SUPPLIER SURVEYS:

SUPPLIER	ASSESSOR	JAN	FEB	MAR	APR	MAY	JUN	JUL	AUG	SEP	OCT	NOV	DEC
ABC COMPANY	BLAKE RIGHT	➡	➡										
DEF CORPORATION	JILL SMITH			➡	➡								
GHI MANUFACTURING	CLYDE JONES				➡	➡							
JKL, INC.	BLAKE RIGHT						➡	➡					
MNP OPERATIONS	BLAKE RIGHT							➡	➡				
QRS ASSEMBLY	JILL SMITH									➡	➡		
TUV SUPPLIES	CLYDE JONES											➡	➡

SELF—SURVEY SUPPLIERS:

THE FOLLOWING SUPPLIERS HAVE BEEN IDENTIFIED BY MANUFACTURING, PURCHASING, AND Q.A. TO SELF-CONDUCT AND SUBMIT A SURVEY (SELF-ASSESSED) DURING THE 1994 CALANDER YEAR. PURCHASING IS RESPONSIBLE FOR REQUESTING AND OBTAINING THE SURVEYS.

☐ ABCDE POLYMERS   ☐ M.J.P. CORPORATION   ☐ F.S.A.B., INC.   ☐ RIGHT CORPORATION

1992 STANDING SURVEYS:

THE FOLLOWING SUPPLIERS WERE SURVEYED IN 1992 WITH GOOD RESULTS. THESE SURVEYS WILL STAND UNTIL 1995 AS LONG AS SUPPLIER PERFORMANCE SCORES ARE 90 OR ABOVE. IN 1995 THESE SUPPLIERS WILL BE PLACED INTO THE 'SELF-SURVEY' CATAGORY.

☐ FGHIJK FABRICATING   ☐ JUNEBUG RUBBER   ☐ MAKING FABRICATIONS   ☐ WHO STEEL, INC.

1993 STANDING SURVEYS:

THE FOLLOWING SUPPLIERS WERE SURVEYED IN 1993 WITH GOOD RESULTS. THESE SURVEYS WILL STAND UNTIL 1995 AS LONG AS SUPPLIER PERFORMANCE SCORES ARE 90 OR ABOVE. IN 1996 THESE SUPPLIERS WILL BE PLACED INTO THE 'SELF-SURVEY' CATAGORY.

☐ MNFG ASSEMBLY   ☐ ABCDE PLASTICS   ☐ WHEN MANUFACTURING   ☐ SHERE CORP.

CUSTOMER SURVEY SCORE SUPPLIERS:

THE FOLLOWING SUPPLIERS HAVE FORWARDED AN ACCEPTABLE SURVEY CONDUCTED ON THEIR COMPANY BY OUR CUSTOMER/S.

☐ MICHIGAN, INC.   ☐ KENTUCKY MANUFACT   ☐ ILLINOIS RUBBER   ☐ CAROLINA FOAM   ☐ WINDI CORP.

☐ SOHIO CORPORATION   ☐ CALIFORNIA, INC.   ☐ NEW YORK METAL   ☐ WASHINGTON SEALS   ☐ TENN METALS

Exhibit 15: 1994 supplier survey schedule (Gantt chart)

SUPPLIER: _____

_____

_____   MONTH _____

ATTENTION: _____   YEAR _____

				SCORE			ACTION REQUIRED
Q U A L I T Y	QUANTITY RECEIVED	QUANTITY REJECTED	% APPROVED=	SCORE	SCORE x .30	1	___ YES ___ NO
	LAST QUARTER S.P.C. DATE RECEIVED ___YES (100) ___NO (0) ___N/A (100)		POINTS=	SCORE	SCORE x .10	2	___ YES ___ NO
	MATERIAL CERTIFICATION RECEIVED _YES(100) _NO(0) _LATE(90)_N/A(100)		POINTS=	SCORE	SCORE x .10	3	___ YES ___ NO
D E L I V E R Y	# OF SHIPMENTS	# ON TIME	% ON TIME=	SCORE	SCORE x .20	4	___ YES ___ NO
		# ASN'S RECEIVED	% RECEIVED=	SCORE	SCORE x .15	5	___ YES ___ NO
		# CORRECT CARRIER	% CORRECT=	SCORE	SCORE x .05	6	___ YES ___ NO
SERVICE AND COST:	PURCHASING'S COMMENTS:_____ _____		RATE 1-100=	SCORE	SCORE x .10	7	___ YES ___ NO

OVERALL RATING=   ADD BLOCKS 1-7 FOR TOTAL=   ___ YES   ___ NO

SCORE MEANINGS:

ANY INDIVIDUAL QUALITY SCORE BELOW 98 REQUIRES A WRITTEN CORRECTIVE
ACTION PLAN. ANY INDIVIDUAL DELIVERY SCORE BELOW 75 REQUIRES A WRITTEN CORRECTIVE
ACTION PLAN. A SERVICE/COST SCORE BELOW 90 REQUIRES A WRITTEN CORRECTIVE ACTION
PLAN. IF THE OVERALL RATING IS BELOW 90, ONE ORE MORE AREAS NEEDS CORRECTIVE
ACTIONS IMPLEMENTED. ACTION PLANS (INCLUDING PREVENTION & VERIFICATION) MUST BE
SENT TO PURCHASING WITHIN 20 DAYS AFTER RECEIPT OF THIS REPORT.

'X'=REQUIRED
'*'=RECEIVED

Exhibit 16: Monthly supplier performance report (log)

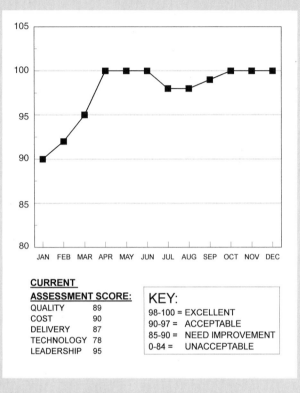

Exhibit 17: Supplier performance
(performance graph)

5   Inspection

5.1   Engineering Standards/Testing

Your company should subscribe to all of your customer's Engineering Standards lists. There are written standards that a Release Engineer may reference on a blueprint that lets you know exactly what specifications the material's properties should meet (i.e., filler content, melt point, flammability, specific gravity, etc.). There are also performance standards for products that measure product life, reliability, and durability. A standard or specification is usually identified first by the organization that developed the standard and then a number representing the standard.

There are many different types of standards used (your company may have developed its own set of standards). Your customers probably have their own set of Engineering Standards or reference another set (i.e., ASTM [American Society of Testing and Materials], ISO [International Organization of Standards], FMVSS [Federal Motor Vehicle Safety Standards], etc.).

If your company assembled a product that consisted of three plastic components, a spring, a self-adhesive foam component, and a pin, your Laboratory Technician might be responsible for the following:

- Obtain Material Certifications from the suppliers of the three plastic resins, spring, foam, adhesive (which may be a certification from the foam supplier's supplier), and pin. These certifications would show what particular Engineering Standard was to be met, what the specifications are, and what the actual measurements of the shipped product was. NOTE: All certifications should be approved by a signature and be no more than one year old.
- Conduct tests on the three plastic components that your company produced and document certification that standards were met or not met (typically you would not certify until the standards were met). These tests might include impact strength, dimensional

stability, weatherability, flammability, color-fastness, and reaction to various chemicals.

- Conduct tests on the finished assembly according to the Engineering Specifications. These might include such tests as life cycle to operate, effort to operate, movement range, humidity and salt-spray.

5.1.1   Certifications and Warrants

Certifications or Warrants are usually required with a submission for production approval and are sometimes required thereafter annually and/or with every shipment of product. The Certification should contain the following information:

- Date,
- customer,
- product number,
- engineering change level,
- product name,
- number or amount of samples tested per the specification,
- specifications and performance requirements,
- results (data plus "pass/fail" notation), and
- signature of the Quality Manager or Lab Technician.

The certification should always be recorded on your company's letterhead, and copies should be retained according to your record retention policy (refer to Chapter 12).

5.1.2   Equipment and Testing Capabilities

The various tests required by Engineering Specifications require special equipment, materials, and fixtures. If your company does not have the necessary equipment, the testing will have to be sourced to an outside testing laboratory. Sometimes your customer may require that you utilize a test lab source that they have approved of. I suggest that if you have to out-source testing, use an ALA (American Laboratory Association) accredited lab.

When using an outside testing lab, you should keep a log that lists what type of test(s) had to be performed. Also list what equipment would have been necessary for you to have conducted the test(s) in-house and what the cost of the test will be. This will allow you to justify equipment purchases in the future when there are evident pay-backs. You should also list three different sources that were contacted to quote the work, what the three quotes were, who was selected, and whether they were chosen for cost, service, quality, or time turn-around.

5.1.3   Useful Documents

As stated above, a list should be kept that details all out-sourced lab testing. That list is kept for reference as well as to serve as a tool in justifying equipment purchases for your company's lab. Another list that should be kept in conjunction with this one should contain information on all tests conducted. It should list the test, standard, product number, customer, date, pass/fail results, etc. If you list the out-sourced testing in bold, you could easily distinguish it and figure the percent of testing that is done in-house each month. That data could serve as a benchmark for improvement in your own lab testing capabilities when establishing departmental goals each year.

A "request form" could be developed for use when any department needs lab testing performed for any reason. This type of form accomplishes many things:

1. It prevents any misunderstandings that can occur during verbal communication,
2. would prevent forgetfulness that sometimes occurs when requests are not put into writing,
3. aids towards the organization and planning of testing activities,
4. would assure that the person requesting the testing provides all of the information that the Lab Technician would need to effectively conduct the test,
5. will equip the technician with a tool to communicate test results, and
6. will serve as a summary of the test conducted and could be retained in the lab.

A request form should ask for:

• Material or product name and number,
• supplier(s),

- specification,
- engineering change level,
- test(s) required,
- requested by,
- date,
- expected report format (formal, informal, verbal),
- date results are needed, and
- any other comments or instructions.

At the bottom of the form could be a section for the technician to summarize the results and whether the sample passed or failed. I suggest also that a small area on the form allows the technician to record the "date promised" and "date delivered." This data will allow you to assess the accuracy of promise dates that the technician establishes so as to continuously improve in that area.

A Lab Manual should be developed that houses or references all lab procedures, policies, unique testing procedures, and equipment use/maintenance. If you do have a separate Lab Manual, make sure that the Quality Manual references it.

An ES (Engineering Standards) Testing Frequencies list should be maintained to record the frequencies of all required reliability, durability, and performance testing. Some parts and assemblies may require testing to occur at given frequencies based on time or volume produced. Such a list will aid in the timeliness and organization of conducting those tests.

Worksheets for frequently conducted tests such as flammability, dimensional stability, filler content, specific gravity, etc. should be developed for lab use. For example, a worksheet for filler content might simply list the material, number, lot number, specification, and date at the top. The worksheet would then prompt the following information in this order:

1. Crucible weight,
2. resin weight,
3. crucible + resin weight,
4. crucible + resin weight after burnout,
5. unburned solids,
6. percent (%) of unburned solids, and
7. average.

Worksheets aid in providing organization of test notes and will serve as reference for the technician on how to conduct the test.

5.2   Inspection Instructions

Inspection instructions should begin to be developed as soon as the blueprint and/or standard is available. Typically, after the Process Flow Chart, FMEA, Control Plan, and packaging specifications are developed and the first sample run has occurred, the in-process/final inspection instructions can be finalized.

5.2.1   Instruction Contents

The instructions should be headed with at least the following information (if applicable):

- Product number and name,
- material specification and your company's code,
- customer,
- blueprint's engineering change level and date,
- gage number,
- color,
- color code,
- container type,
- quantity per container, and
- number of cavities.

Not all of this information would be applicable to all types of product, but it gives you an idea of the instruction heading contents. Any other type of nomenclature that is unique to the product or its material(s) that your company uses for identification should be listed as well.

The Acceptance Standard Criteria should list all inspections that are required by the customer and your company. Each criteria should state the method of inspection (i.e., visual, using check pins, with calipers, tensile tester, etc.) and the specifications (i.e., 6 mm +/- .3 mm, no more than .5 mm of parting line flash, etc.). If visual inspections state such criteria as "no *heavy* flow lines," it should also state "please refer to master sample or accept/reject visual board for determination of acceptability." To check and record the raw material(s) lot number should *always* be listed under Acceptance Standard Criteria to sustain traceability.

All criteria that is to be recorded on an SPC chart should also be identified as such. Criteria that is considered "critical" (i.e., marked as such on blueprint or characteristic is very important to the next operation) should be identified with a symbol or "C" or otherwise labeled as being "critical." Any safety-related criteria should also be identified with a symbol or "S" to signify that it is a safety inspection. All other criteria should be considered "significant."

It is wise to include an extra page to the instructions that contains a drawing of the product (if applicable) so that inspection criteria can be referenced to it. That will enable the inspector to easily identify which hole is to measure 1.3 mm, or where the unacceptable flow line is most likely to occur, for example.

After the inspection criteria are listed, there should be detailed packaging criteria written on the instructions to ensure an effective check of the packaging (i.e., right hand and left hand goods must be packed separately; nest 2 back-to-back and pack 36 in a row, 5 rows high; 4 corrugated dividers must be used between rows. Or, bulk package 500 pounds in plastic-lined gaylord, etc.).

A Rejection Summary should appear on the inspection instructions and inform the inspector to refer to the Rejection Procedure for further details. It is very important to state in this area that "Zero discrepancies is the basis for accepting or rejecting." This means that if even one defect is found, the entire population represented would be rejected.

Some companies prefer to have separate instructions for each phase of inspection. I have found it to be less confusing to develop one set of instructions and simply include a summary of the inspection frequency for each phase. For instance, after listing all of the criteria, a section could be added to the instructions:

- *Start-up inspection*:   Inspect one (1) complete sample to all items listed under "Acceptance Standard Criteria" when the job is started. Follow the "Start-up Inspection Procedure" for details of approval and the posting of the first sample.

- *In-process inspection*:   Inspect one (1) complete sample to all items listed below "Acceptance Standard Criteria" once every 2 hours, and record results on the inspection log sheet in the job file. Follow the "In-process Inspection Procedure" for further details.

- *Final inspection*:   Refer to the "Lot Acceptance Sampling Table" for the quantity to be inspected in each finished lot. The criteria that requires SPC charting need not be inspected. Stamp container labels as "Approved" to accept, or tag as "Rejected" and follow the

"Rejection Procedure" if a discrepancy is found. Record the rejection serial number on the log sheet.

Details of the phases of inspection will follow in this chapter.

When the inspection instructions are completed, record the issue date and number (i.e., "release," 2nd issue). They should then be approved and signed by the Quality Manager before release (refer to Exhibit 18 for an example of completed instructions).

5.2.2   Lot Acceptance Sampling Table

Please refer to Exhibit 19 for an example of a statistically accepted "Lot Acceptance Sampling Table." I referred to this as part of the instructions for Final Inspection. The table would not change for each set of instructions but could be used in conjunction with all Final Inspection Instructions. Basically, the inspector would locate the amount of product in the lot on the left-hand side of the table. They would then follow the table to the right and inspect the amount indicated for safety items, critical items, and significant items.

5.2.3   Maintenance and Location of Instructions

I suggest that all inspection instructions be developed on computer so that updates, improvements, and revisions can be made quickly. Once on computer and printed, the original printed instructions should be placed in a "Master" binder and kept by the Quality Manager. The "Masters" can then be used to make copies whenever needed. There should always be a copy of the instructions displayed at the operation when running, and in the job file. You may find it beneficial also for the inspectors to have their own binder where copies of the instructions can be placed when a job is running. Their log sheets used to record inspection results could be placed facing the instructions in the same binder.

5.2.4   Instruction Improvements and Revisions

If the discipline of treating the inspection instructions as true "living documents" is instilled in the Quality and Manufacturing departments, I see no need for any formal reviews of them. If indeed the instructions are reviewed, updated, and improved as a result of normal processes, (refer to the next two paragraphs), a formal annual review would be futile.

Revisions and updates to the instructions should be prompted by Engineering Changes, Packaging changes, etc. (i.e., the material, testing, part number, slot size, container type, color, etc., changed). When revisions/updates are prompted through whatever means (usually an Engineering Change Notice), the Quality Manager would enter those changes/updates on the instructions and reprint a "Master." The obsolete instructions should immediately be removed from the system and replaced with the next issue.

There are a lot of other events, problem solving efforts, and quality improvement efforts that should also provoke changes and/or additions to the current inspection instructions. If, for example, during problem solving for a particular product concern, the team decides that an impact test should be added to the in-process inspection to prevent reoccurrence of unacceptable product being shipped, then that process should be detailed on the instructions. Throughout the duration of a program, it is very possible that standards will be developed verbally with the customer and should also be documented on the instructions. For example, a visual defect appears on a product as a result of a process change that was intended to improve the dimensional integrity of the product. The Quality Manager contacts the customer because the seriousness of the defect is questionable in regard to its nonvisible location in an assembly. The customer receives a sample of the product in question and relays to the Quality Manager that the defect is acceptable. The Quality Manager would definitely include with the current inspection criteria that the defect was acceptable.

Another example of when to improve instructions would be when a quality improvement or process optimization occurs that affects the current criteria. For instance, a Design of Experiment was conducted to improve part quality and it was found that only virgin material should be used to accomplish the quality desired. The Quality Manager should make sure that a criterion is added to the instructions that states "parts are to be manufactured with virgin material only."

5.2.5 Procedure

Once your company has decided on the format of your inspection instructions and the process in which they will be developed, revised, improved, maintained, and used, a formal procedure for their development should be documented. All quality-related procedures should be placed or referenced in the Quality Manual.

5.3   Start-Up Inspection

Whenever a manufacturing process is started up or has changed (i.e., different material being used, blocked off one cavity, part of the process went from an "automatic" mode to a "manual" mode for some reason, the color changed, etc.), Quality should approve the product to allow the process to continue. A lot of injection molders and blow molders will submit a "first shot" to Quality, while extruders and processors usually submit a weighed quantity or sample from the beginning of the "first batch." Understand that "first shot" does not necessarily mean that the sample is actually the first shot produced.

5.3.1   Starting Up the Process

Typically, the foreman or whoever is responsible for starting up the process will set up the machine parameters according to the process set-up instructions (refer to Chapter 8 for details on process set-up development and deviations). He/she will then allow the process to cycle a few times while visually inspecting the product. Adjustments to the normal set-up may be made within specified tolerances during this time to compensate for changes from the last run (i.e., different lot of raw material/s, different weather that affects the process, different machine, new operator, etc.). When the Foreman believes (according to the process set-up and all visual requirements) that the process is producing an acceptable product, then he/she would submit the product to Quality.

5.3.2   Approval of the Process

When Quality receives the first piece/shot/sample from manufacturing, it is usually the Inspector who checks the product according to the inspection instructions. After all gaging, testing, measuring, visual inspection, etc., is complete, the product is deemed either acceptable or unacceptable by Quality, labeled as such, and returned immediately to manufacturing. If the product is unacceptable, all goods produced thus far and thereafter should be rejected until the process is approved. If the product is acceptable, the Inspector should return the sample to manufacturing with the following information attached:

- Date,
- time submitted,
- time approved, and
- inspector's name.

The approved "first shot" should remain at the operation through the duration of the run for reference and later first/last sample comparison reviews. Any product produced prior to approval of the process should be either rejected or verified for approval.

5.3.3 Procedures and Instructions

A procedure that describes your company's process of start-up inspection should be documented and included or referenced in the Quality Manual. Inspection instructions for start-up inspection should list all criteria and standards that the product must meet and should be unique for each product.

5.3.4 Helpful Hints

- If your process produces molded products, there may be date codes on the product that should be changed in the tool at certain frequencies. If this is so, it is wise to have the Inspector circle the date code with a grease pencil to signify that it is up-to-date. Also include the date code check on the inspection instructions and in the procedure.
- Try to have the approval information (date, time submitted, time approved, inspector name, etc.) be applied to a label with an adhesive backing. This type of label will accomplish three things:

1. It will remind the Inspector to fill in all of the information required,
2. It will make it easier for the Inspector to attach the approval label to the product or container, and

3. It will be less likely to become removed from the start-up sample or lost.

• If possible, have your procedure include that after the Inspector has approved the product, the Supervisor should initial the approval label. This is particularly effective if your product is decorative or visual because it gets an extra pair of eyes to potentially spot a discrepancy that may have been missed by the Inspector.

5.4   Process Inspection

After the process has been approved by Quality and the "accepted" first-sample is posted at the operation, inspection should continue at specific frequencies throughout the duration of the run. Inspection should be conducted in accordance to your Inspection Instructions Acceptance Standards.

5.4.1   Establishing Frequencies

The frequency of inspection will depend on many different factors. It makes sense that if a particular inspection requires a test that takes four hours, it wouldn't be conducted more than once every four hours unless, statistically, it has been determined that failure will likely occur more often (refer to your FMEA's occurrence ratings). Inspection frequencies also depend on where your company is in regard to progress with SPC, Process Control, DOE, etc. The more advanced a company is with these analytical tools and controls, the less they will have to be used. In other words, the Inspector would spend more time detecting potential problems and getting them corrected than detecting problems after they have occurred. Of course, the difficulty and cost of certain types of inspection will aid in determining frequencies as well. No one wants to pay $50.00 for a plastic product that underwent expensive inspection when they could pay $10.00 for the same from a different supplier who has control over their processes.

   In-process inspections could occur as often as every half-hour or as infrequently as once per shift. For example, if you are producing plastic products, an initial in-process inspection program might require that SPC is performed every two hours at each machine and all required dimensional and visual checks are performed on one current sample every hour. As control of the processes becomes evident through SPC as a result of process optimization, a company may decide to cut product inspection in half. The time gained by the Inspector could then be utilized in effective preventative problem solving, DOEs, and other preventative types of monitoring rather than detection.

5.4.2 Inspection Records

Any time inspection occurs, it should be documented. Results of inspection should be logged by product number, not by the date of manufacture or the machine/process. When developing a log sheet, make the format as simple as possible so that it is very evident to the user exactly what is to be recorded and where. For example, the top of the log form should contain spaces (and labeled as) "Customer," "Product number," "Calendar year." Horizontal and vertical lines can be drawn to form rows that can contain the inspection information and columns that can be headed as the following: "Date," "Time," "Initials," "# Inspected," "# Accepted," "# Rejected," and "Material lot number." The rest of the columns can be headed as "Item 1," "Item 2," etc., to correspond with the acceptance standard criteria that is listed on the Inspection Instructions. In these columns, the inspector would record the *number* of parts rejected for not meeting the criteria. I suggest also, if the inspection requires an actual measurement to be taken that is not recorded on an SPC chart, the actual range of measurements be recorded in those columns. The last column should be reserved for "Comments" so that rejection serial numbers, deviation numbers, shift-to-shift communications, etc., can be documented.

The inspection records should be located with the inspection instructions while the job is running and filed by product number when not in production. They should be retained beyond that for at least two years (five years if the product is controlled by government safety regulations).

5.4.3 Procedures

Make sure that there is a formal procedure located in or referenced in the Quality Manual that describes exactly how the inspection log sheet is to be filled out and how the records are to be maintained. Inspection records are quite simple to keep, but because they are so critical to your company's record archives, they should be regarded with care, enforced, and reviewed regularly by the Quality Manager.

In addition to, or combined with, the Log Sheet Use Procedure should be procedures written that describe your company's In-Process Inspection System. It should include the purpose and scope of the inspection as well as what documents are used and who is responsible. It should also be specific in regard to how much is inspected and how often.

5.5   End-of-Run Inspection

At the end of a run, the foreman or whoever is responsible for shutting down the operation can submit the final "shot" or last batch sample to the Quality Inspector. In Section 5.3 entitled "Start-Up Inspection," I mentioned that the first product sample be retained at the operation for reference during the run and would be useful after the run for comparison to the last sample. That sample should also be removed from the operation at the end of the run and retained by manufacturing. Once Quality has inspected the last sample to the same criteria required of the first, they should forward the sample back to manufacturing. The sample should be labeled as "last sample" and the label should include the run dates, machine number, indication of whether it was "accepted" or "rejected," and the inspector's initials.

5.5.1   Documentation of Inspection

I suggest that a special "First/Last Inspection" log sheet be made up specifically for these inspection records to list the data and information for *both* inspections. For easy comparison purposes, the first inspection records could be located down the left-hand side of the sheet and the last inspection records down the right-hand side. At start-up, the log sheet could be posted at the operation along with the first sample and returned by manufacturing at the end of the run with the first and last samples. At this time, the inspector would conduct the exact same inspections and tests on the last sample and log the results beside the first sample's results. At the bottom of the log should be a space for the Quality Department's and the Manufacturing Department's recommendation for review, and Action Plan.

5.5.2 Initial Evaluation of Results

After the last inspection is complete, the Inspector would decide whether the product was acceptable or not. If not, the lot produced should rejected, quarantined, and handled according to your rejection procedures. If it is acceptable, the lot should continue through the normal inspection and storage processes. The inspector should then sign the report, date it, and make a recommendation for formal review (see "Recommendations for Review" below) if they think it necessary. After the log sheet is completed to that point, a copy should be handed back to the Manufacturing Foreman or Supervisor with the last sample from the run (don't forget to retain the original log in the part file). At this time, Manufacturing can review the results and make a recommendation for review if needed.

5.5.3 Recommendations for Review

Recommendations for a formal last sample review should be made by Quality if the part or product was unacceptable, displayed substandard quality, etc. For example, if a hole size was out of specification, the appearance was borderline, or the filler content was on the high side, Quality would recommend a formal review of the first and last samples. If the operator had to trim flash, sort out a high percent of shorts, or run the process in an abnormal fashion in order to produce an acceptable product, Manufacturing would recommend a formal review.

5.5.4 Formal Review—Comparison of First and Last Samples

If a formal review has been recommended by either Quality or Manufacturing, a Manufacturing representative should bring the first and last retained samples and the inspection record (log sheet) to a meeting where they can be reviewed.

Your company may decide to hold a meeting specifically for the purpose of comparing first and last samples from each run and reviewing the results. I suggest, however, that it is probably not necessary, and an acceptable review of the samples and data could occur during an already scheduled daily or weekly meeting. You may decide that an already established Engineering, Accountability, Quality, etc. meeting can conduct the formal review.

It should, however, be a meeting that is multi-disciplined for optimum benefits.

During the review, the following should be discussed:

1. First and last sample comparison (major differences in data would indicate significant variation in the process and should be analyzed),
2. reason for recommendation of review (a concern that is documented on the report by Quality and/or Manufacturing), and
3. corrective/preventative action plan (developed by the team with responsibilities and expected due dates assigned).

5.5.5   Helpful Hints

If your operation is a such that changeovers are time consuming, costly, etc., you may want to make it policy that the process cannot be changed over until the last sample has been submitted by the Foreman and approved by Quality. This practice would alleviate the need to re-set an operation after it has begun changeover as a result of Quality later finding the last sample to be unacceptable.

5.5.6   Procedures

This process, once developed by your company, should be documented into a procedure and included or referenced in the Quality Manual.

5.6   Final Inspection

All product should be subjected to a final inspection once it is packaged and ready to be moved to finished goods or an in-process operation. This is extremely important because there are many factors that can influence the quality of the product after it has been produced. For example, during an in-process inspection of a current shot or sample taken from the operation, it is impossible to know whether or not the operator is correctly trimming a gate, packing the product correctly, consistently operating the process, etc., after the Inspector walks away.

If your product could in no way be affected by other factors after an in-process inspection occurs, then final inspection is not needed. It is important to note here that a sufficient quantity from the lot should be inspected according to the Lot Acceptance Sampling Table (see section 5.2, "Inspection Instructions") before a container of a product is approved.

5.6.1   Procedure

Final inspection of a lot should involve all checks and tests that are required under "Acceptance Standard Criteria" on the Inspection Instructions sheet. If you are conducting SPC on some measurements that in no way could have changed or been influenced by factors after the in-process inspection, those measurements would not have to be repeated during this phase of inspection. Also, if you have inspected a sufficient quantity (according to your final inspection sampling table) from the lot during in-process inspection, any other inspection results that you know will be unchanged do not have to be re-inspected during final inspection. If your company wants to practice final inspection with this rule, it should be very careful that, indeed, the product could not have changed.

Aside from the Acceptance Standard Criteria, special attention should by paid to packaging (especially when the process is producing more than one part number), traceability identification of the contents, and labeling during final inspection. Inspection of container labels should be a "cut and dry" process and (in my opinion) the simplest part of final inspection. On the contrary, it is usually one of the top three reasons for customer rejection/ complaint and causes many problems for our customers in both inventory control and in the use of wrong material/products during their operation.

5.6.2   Frequencies/Quantities

Final inspection could occur for every container, batch, or lot. It is a decision to be made by your company depending on its needs. The important point here is that a sufficient quantity be inspected according to a Lot Acceptance Sampling Table similar to the one located in Section 5.2, "Inspection Instructions."

5.6.3   Inspection Records

Detailed records of the final inspection should be documented and retained in the same manner that "process inspection" records are. Please refer back to Section 5.4.2, Inspection Records for examples.

5.6.4   Helpful Hints

It should be company policy that no labels are to be placed on an empty container! Labels should be placed on containers *after* they are packaged with the product. A label on an empty or partial container will inevitably result in mislabeled containers. Prelabeling also enables an Inspector to approve the container before the product has even been packaged (I know this would never happen at your plant!). Also for obvious reasons, never let the person who labels the container inspect and approve their own work (label).

5.7   Dock Audit Inspection

Dock audits can provide excellent information in regard to the effectiveness of your company's inspection process and can also serve as a useful tool to direct continuous improvement efforts in the inspection system.

A dock audit process can be constructed in many various ways. Basically, a dock audit is a full inspection of a container from a production lot of *all* criteria listed on the inspection instruction sheet. The chosen container (usually at random) will represent the entire lot. The results of the audit will be assumed to be the status of the entire lot. This means that if the container meets all inspection criteria, the lot meets all of the requirements. The same assumption is expected if the chosen container was found to be unacceptable for some reason—the entire lot would be assumed to be unacceptable and therefore would be rejected according to your procedures.

5.7.1   Types of Dock Audit Processes

I mentioned that this audit process can be constructed in many various ways. The following examples are some of the more typical procedures in use:

1. A carton is selected at random from every lot of production for an audit. A "lot" can be defined as:
   A. A production run,
   B. the production from a 24-hour time period, and
   C. the production from an 8-hour time period. A "lot" also consists of the same product identification number and/or material.
2. A carton is selected at random from a lot only when one or more of the following conditions occur:
   A. The scrap rate during production exceeded a standard or was out of the control limits on an SPC chart.
   B. There was a new or inexperienced employee responsible for the quality being packaged.
   C. The product falls on a monthly or weekly chart that lists either the top customer quality concerns or internal concerns.
   D. The product had to be reworked or sorted for previous quality concerns.
3. A predetermined number of dock audits are expected to occur within a given time frequency.

It is important to consider the *needs* of your facility in addition to available manpower when determining the type of dock audit process you may incorporate into a quality system.

5.7.2   Who Should Perform Dock Audits?

Your company's size, production output, and time required for your plant's dock audit procedures to be properly followed will be major determining factors as to who should be responsible for performing the audits. The very important factor to be considered is that a dock audit should *never* be performed by the operator or inspector who conducted the in-process or final inspections on the product. The reason for this is because one of the purposes for these audits is to inspect the inspection system and it is difficult for someone to "catch" their own mistakes. If your company is large and opts for an extensive audit process, there may be need for a full time Dock Audit Inspector. Maybe those responsibilities could be combined with others (i.e., quarantine maintenance, an inspector from a different department, shipping department hand, etc.) and simply added to their job description. A very creative way to accomplish a smaller scale dock audit process could be to assign one per day/week to salaried personnel on a rotating basis. Not only is this creative but it would provide an excellent opportunity for the "front office" personnel to become familiar with the company's product. This type of assigned responsibility would assure useful feedback in regard to the effectiveness of inspection instructions. Also, "fresh eyes" inevitably will question quality differently than those dealing with that type of decision making on a daily basis.

5.7.3   Documentation

I suggest that a preprinted dock-audit form be constructed to be used by the auditor. The form should contain spaces at the top that require the following information be recorded:

- Product identification number,
- product description,
- customer,
- audit date,
- auditor,
- date of manufacture (and shift number if applicable),
- container amount,
- operator (this should be traceable from the label or packing slip), and
- identification of the final inspector.

The center of the form could depict a table with columns and rows. The columns could be headed with the following:

1. Inspection criteria,
2. total amount inspected, and
3. amount rejected.

Along the top of the table, the row headings can be filled in by the auditor by copying the "Acceptance Standard Criteria" from the inspection instructions. The amount inspected and amount rejected for each criteria then can be recorded in each column beneath the criteria.

The bottom of the form could provide spaces for the overall percent rejected, and whether or not the entire lot has been deemed acceptable or rejected. There should also be space allotted to record corrective actions if indeed the lot had to be rejected.

Whenever a dock audit is performed and the paperwork is complete, it should be filed in the product file in Quality.

5.7.4   Continuous Improvement Opportunity

To assure continuous improvement in your inspection process, the data from your dock audit process can be utilized. I suggest that your company schedule a meeting every two weeks or every month following your regular quality meeting, etc., to review all dock audits that resulted in a rejection. There should, at a minimum, be representatives from Quality and Manufacturing along with whoever is responsible for conducting the audits at the meeting.

When reviewing audits that resulted in a rejection, the team should decide if the discrepancy *should* have (according to the inspection procedures) been detected by the Inspector(s). In some cases, there may have been one unacceptable product in the bottom of a container of 2,000 pieces. Because an Inspector would typically be responsible for only a certain percent of the contents, the team would determine that the inspection process failed at the operation when the operator inspected and packed the part. If 100% of the product audited was found to be discrepant, the team would undoubtedly realize that the in-process *and* final inspection conducted by the Inspector failed.

Once the team has agreed on *where* the inspection process failed, they should begin to develop a corrective action that would prevent that type of failure from ever occurring again. Don't confuse this part of the dock audit process with the corrective action/s that should occur immediately when product is rejected. The preventative actions developed in this meeting should be focused on the *inspection process*, not on the event that actually *caused* the product to be defective. The preventative action plans should be documented via meeting minutes or individual reports with the name of the person responsible for the action and the date it is anticipated to be implemented.

As a measurement tool to provide the team with proof that the continuous improvement process is (or is not) working, a chart could be constructed and updated at least monthly. It would show *how many* audits were conducted each month, how many resulted in rejections, and the *percent* that resulted in rejections. The percent should be displayed in graph form (bar graph, line graph, etc.) so that meaningful trends, freaks, etc. can be evaluated.

If this continuous improvement tool is used seriously, it can result in major improvements of your company's inspection process. Ultimately, it is possible that eventually your company will experience periods of six months or longer without any dock audits resulting in lot rejections. When this type of success is evident, you may decide that your dock audit procedure can be changed to require less audits than before.

5.7.5   Helpful Hints

Whenever a dock audit results in a rejection, it is a good practice to copy the completed form to both the inspector and the operator listed on the packing slip. This communicates to them that there was a failure in the inspection process and enables them to contribute information that may be helpful in preventing a recurrence of the same problem.

It is also encouraging to operators and inspectors to see a completed audit form that shows that the container they were responsible for was found to have no discrepancies. You could even write "Good job!" across the top of their copies.

5.8   Regrind Control

This section is meant primarily for anyone whose company grinds plastic already processed to be reused for the same or another process. Examples of this would include the grinding of runners from an injection molding process, tabs from a blow molding process, scrap from a stamping process, rejected and obsolete plastic products, etc. By "Regrind Control," I intend to stress the importance of the identification, problem prevention, and traceability of regrind materials. In regard to how many generations regrind material may be used and what percentages of regrind can be mixed with virgin materials to produce acceptable products, specifications should be based on expectations and developed prior to production by you, your supplier, or your customer for each individual product.

Too often, regrind material finds its way into a product that should have been made of a different material or into a container of virgin material, only to contaminate the entire amount because it was a different type than the virgin. To eliminate or reduce the occurrence of these (and many other) types of disasters, a control system would have to be put into place.

5.8.1   Control System

The control system could begin by documenting requirements on inspection instructions and on material handler's instructions (i.e., "product must run out of 100% virgin material," "no more than 20% regrind allowed," etc.). A process should be outlined and formally documented for Material Handlers to properly adhere to the requirements of each job. In other words, if a job is allowed no more than 20% regrind, studies should have already been performed that determine runner/tab weight ratio to shot weight, in addition to the amount of product that can be ground out of a certain amount produced. The results of such a study would provide the information needed to know whether or not material ground at the operation could immediately be mixed with virgin at given frequencies or stored for later mixing. It is this type of instruction that should be determined and documented so that the Material Handler knows how to (and is accommodated to) effectively perform the job.

If regrind is being mixed with other material, the process should be set up so that mistakes are not easily made. For instance, if it has been determined that all requirements of a job can be met, even with the regrind of runners and scrap being circulated back through the process, then you might want the grinding operation placed at the work station. To further protect product quality, you may even have regrind mechanically transferred from a grinder to a machine hopper while mixing with transferred virgin material from a different container. After assuring that hoses are connected properly and to the right machine, grinder, and material container, it will be almost impossible to contaminate another material with the regrind.

Establishing a work flow that accommodates problem prevention is important. There should be documentation of specifications (on inspection and material handling instructions), examination of incorporating adherence to specifications into the overall process, documentation of the overall process into a procedure, training of those who are expected to follow the procedure, and follow-up.

5.8.2   Traceability

All materials should be identified at all times with the following information:

- Material type,
- color,

- color code,
- supplier identification number,
- your company's nomenclature (if applicable),
- date produced, and
- amount of material in the labeled container.

A system should be implemented that allows traceability to the purchased or manufactured material. Therefore, if a purchased or manufactured virgin material (or even a reprocessed material) is ground, it should be relabeled with all the above information, in addition to how many times the material has been ground and the date it was relabeled for use. The ground material should also be clearly identified as such or as "Regrind."

5.8.3   Regrind Approval for Use

I suggest also that all regrind be approved for use by Quality. Inspection of regrind may be as simple as a flame, smell, and/or visual examination to properly identify the material. As well, inspection could go as far as to involve the testing of the material's properties for conformance to the blueprint specifications. It is very important that, at a minimum, Quality assures that the regrind is properly identified/labeled correctly.

5.8.4   Helpful Hints

If regrind material is mixed with virgin material away from the process or for storage, the label should indicate what percentage of the material is virgin and what percentage is regrind. This will take the guess work out of the material handler's job when he/she provides material that is to meet a certain specification of regrind percent allowable.

Include in your procedure that when material is relabeled because it has been ground or had regrind mixed with it, all other labels be removed from the container. This should also be included in the Quality inspection instructions and be mandatory before the container can be approved for use.

5.9   Color Control

If your company produces products that are made out of colored materials and/or materials which color must match a "master," a color control program should be documented to assure acceptable products. The color control process should include received, manufactured, submitted, and to-be-sold products that are intended to meet color specifications.

5.9.1   Inspection Phases

Color matches and procedures for color approval should be incorporated into all inspection phases that have been discussed so far in this book. Those phases include:

1. Production sample runs,
2. sample submission to the customer or production approval source,
3. pilot runs and submissions,
4. receiving inspection,
5. start-up inspection,
6. process inspection,
7. end-of-run inspection,
8. final inspection, and
9. dock audit inspection.

5.9.2   Color Inspection Tools

If any of your products require a 100% color match, usually a MacBeth light or similar multi-type light unit is necessary for proper inspection. The actual specification that requires 100% color match should describe exactly what criteria the tools to be used should meet. Some criteria may even state that the color should be checked mechanically via a spectrophotometer. Another tool that may be required is a gloss meter. Gloss is very important when checking colors because it can significantly affect the color appearance of a surface in the "light/dark" range.

The most frequently used tool is the human eyes (mostly in conjunction with the MacBeth light, sunlight, or fluorescent light). Just as any other inspection tools should be certified as to accuracy, so should the eyes that your company will empower to deem a product acceptable or unacceptable for color match. There are many products and services that your company can purchase to evaluate the ability of your inspectors and operators to inspect colors. I have found that the Munsell Hue Test package by MacBeth meets most expectations in regard to quality, cost, and ease of use for testing one's color inspection abilities. I urge you to seek and tap into the many available sources and have a team decide on the product or service your company will use. Do not blindly assume that the product I have found to be acceptable for my company will be the best for yours.

5.9.3   Helpful Hints

Color "masters" should be maintained in a dust-free container (zip-lock plastic bag) and never be handled except with gloves. Dust and oils from human skin can quicken the aging process of the masters and distort the original color and gloss.

When a color is changed in a process, a new "first shot" should be inspected and replace the previous color that was placed at the process.

If you request on your purchase orders that a color chip representative of a shipped lot of precolored material accompany the shipment, approval of those materials in Receiving Inspection can be accelerated.

TOOL NUMBER	DESCRIPTION	CUSTOMER	ENGINEERING LEVEL	BLUE PRINT DATE	PART NUMBER	
400200 400300	TRIM INSERT	XYZ CORP.	#52 DATED: 05-JA-90	20-NOV-89	12345678 (RIGHT HAND) 12345679 (LEFT HAND)	

MATERIAL CODE	MATERIAL SPEC.	COLOR	GAGE NUMBER	CAVITIES	CONTAINER TYPE	CTN. QUANTITY
01XXXX0000	ASTM-ABS 002	NATURAL	400300	2 # CAV'S 1 R.H. 1 L.H.	HALF GAYLORD	130 PCS.

ACCEPTANCE STANDARD CRITERIA

1. NOTE: RECORD MATERIAL AND LOT NUMBER ON INSPECTION LOG SHEET. INSURE THAT THE CORRECT RAW MATERIAL SPECIFIED ABOVE IS IN THE GAYLORD AND HOPPER PRODUCING THESE PARTS.

2. PART NUMBERS ON LABELS, PARTS, AND INSPECTION INSTRUCTIONS MATCH. PROPER ENGINEERING LEVEL ON LABEL. NO CARTON IS TO BE APPROVED WITHOUT A "PACKING SLIP" INCLUDED.

3. FIRST PIECE APPROVAL POSTED AT MACHINE: ACCEPTABLE TO "MASTER" AND DATE CODE ON PART IS CORRECT.

4. VISUAL: ABSOLUTELY NO SHORTS, FLASH, BROKEN PINS/STUDS, OIL, MOLD SPRAY, SPLAY, HEAVY FLOW LINES, PULLS.

5. VISUAL: NO EXCESSIVE WARPAGE IS ALLOWED. CHECK LENGTH AND CONTOUR TO GAGE (SEE GAGE # ABOVE).

6. CHECK PEGS (6) W I D T H WITH CALIPERS. SPEC. = 4.6 TO 4.8 mm
   -PLEASE REFER TO PRINT FOR MARKED LOCATION.

7. CHECK PEGS (6) H E I G H T WITH CALIPERS. SPEC. = 9.5 TO 10.5 mm
   -PLEASE REFER TO PRINT FOR MARKED LOCATION.

8. SPC: SEE SPC FOLDER IN JOB FILE-
   FOLLOW INSTRUCTIONS FOR GAGING, FREQUENCY AND SAMPLE SIZE TO MEASURE.

9. IMPACT: MUST NOT CRACK AT 60 IN/LBS AT AREAS NOTED IN SPC FOLDER. LOG PASS/FAIL RESULTS IN IMPACT LOG. USE GARDNER IMPACTOR LOCATED IN Q.A. LAB.

10. STAPLE: MUST NOT CRACK WHEN STAPLED AT LOCATION POINTS NOTED – PLEASE REFER TO PRINT FOR MARKED LOCATION. USE HEAVY DUTY STAPLER LOCATED IN Q.A. LAB.

INSPECTION INSTRUCTIONS:	REJECTION SUMMARY:	PACKAGING:
IN-PROCESS: INSPECT 1 COMPLETE SHOT TO THE ITEMS LISTED BELOW "ACCEPTANCE STANDARD CRITERIA" ONCE EVERY 2 HOURS. RECORD RESULTS ON INSPECTION LOG SHEET FROM JOB FILE. **FINAL:** REFER TO THE CORP. QUALITY & PRODUCTIVITY LOT ACCEPTANCE SAMPLING TABLE. STAMP LABELS WITH "OK" TO APPROVE. FOLLOW REJECT PROCEDURE IF ANY DISCREPANCIES ARE FOUND AND RECORD SERIAL # ON REJECT TAGS AND IN "COMMENTS" SECTION ON THE INSPECTION LOG SHEET. **DOCK AUDIT:** INSPECT 1 CTN. 100% TO ALL VISUAL ITEMS AND 10% OF ALL DIMENSIONAL ITEMS OF "ACCEPTANCE STANDARD CRITERIA" AND RECORD RESULTS ON A "DOCK AUDIT" FORM. 1 RANDOM CTN./EVERY LOT.	ZERO DISCREPANCIES IS THE BASIS FOR ACCEPTING OR REJECTING. **REJECTION PROCEDURE SUMMARY:** PLACE A COMPLETELY FILLED OUT ORANGE REJECT STICKER ON CARTON/S. ASSURE THE CTN.'S ARE REMOVED FROM PRODUCTION FLOW AND HELD IN QUARANTINE FOR THE DISPOSITION. FOLLOW REJECT PROCEDURE: FILL OUT FORM, PLACE REJECTION INFORMATION AND SERIAL # ON INSPECTION LOG SHEET AND REJECT STICKERS. GIVE HARD COPY OF REPORT TO FOREMAN AND PLACE QA COPY IN TRACKING FILE BOX. SEE REJECT PROCEDURE FOR MORE DETAILED INSTRUCTIONS.	RH AND LH PARTS MUST BE PACKED SEPERATE! HALF GAYLORD. ACROSS BOX WIDTH. 4 ROWS OF 75 PCS, ON SHORTEST EDGE WITH FLANGE. PLACE 12 STACKS OF 5 PCS ON TOP. RADIUS UP. TOTAL PIECES: 360 SEE S.O.P. BOARD FOR VISUAL AID

ISSUE NUMBER ___5___   ISSUE/REVISION DATE ___01-12-94___   APPROVED BY _J.R. Waterman_

Exhibit 18: Inspection instructions (in-process, final, and dock audit)

LOT OR SHIPMENT SIZE	SAMPLE SIZE PER CHARACTERISTIC CLASSIFICATION		
	SAFETY	CRITICAL	SIGNIFICANT
0 - 15	100% INSPECTION		
16 - 25			
26 - 50			
51 - 75		50	35
76 - 125	75	65	40
126 - 225	90	75	45
226 - 445	100	85	45
426 - 1300	110	90	50
1301 +	115	90	50

Exhibit 19: Lot acceptance sampling table

6   Gage Control

6.1   Gage Fixture Certification

All gages and holding fixtures should be certified to all required construction tolerances before they are used. It should be mandatory for the gage source to have verified all specified dimensions as required by the blueprint before submitting it to its customer. The certification of a gage or holding fixture goes beyond the initial verification of the gage source.

6.1.1   Third Party Certification

Third party certification is conducted by an independent layout or metrology service. I recommend the use of an independent third party certification because it is conducted by a source that will be 100% objective on reporting the results. This is critical when your company's layout technician, inspectors, engineers, etc., will assume that all data gathered with the assistance of the gage or fixture is reliable.

If your customer requires that an independent third party certification be conducted, find out if they also require that you use one of their approved sources. If so, obtain their "approved source" list for gage/fixture certification or metrology/layout and adhere to it.

I recommend that your company include, in all purchase orders with your gage and fixture build sources, that they are responsible for obtaining an independent third party certification. The purchase order should also state that any gage or fixture not meeting required construction tolerances be corrected before it will be accepted by your company.

6.1.2 In-House Gage/Fixture Verification and Acceptance

Once a gage or fixture is completed, certified, and submitted to your company, it should be verified by metrology or layout. Upon receipt of the gage/fixture, the layout person should audit and approve the source's third party certification as being accurate with engineering specifications, (and certified by a customer approved source if applicable). Layout should also ensure that the gage/fixture is to the latest design level and enter the I.D. number into your company's calibration and gage repeatability and reproducibility (R&R) programs (continue in this chapter for details on calibration and R&R).

Approval could be indicated by placement of a company calibration sticker onto the gage or fixture which shows the due date of the next calibration. If all of the verification and acceptance criteria is not met, your rejection procedure should be followed and the gage or fixture returned to the source.

6.2 Gage Repeatability and Reproducibility (R&R)

Repeatability and reproducibility studies should be conducted on all gages at the time of receipt and at least annually thereafter. If a gage or the gaging method is modified in any way, or new operators/inspectors will be using the gage, a new R&R study should be conducted. It is very important that the person/s who will be using the gage be the appraisers during the study. The process of obtaining production approval on a product should include the requirement of an acceptable gage R&R study. If persons other than the actual appraisers are used to conduct the study, it could cause major problems when the product goes into production. The measurement error may be unacceptable due to poor training of the appraisers or poor instructions for the gaging method. This could result in acceptable product being rejected based on inaccurate data. Even worse, unacceptable product may be approved to be shipped to the customer.

6.2.1 Short-Form and Long-Form Studies

Typically, to conduct a gage R&R study, three appraisers would each take measurements from a group of ten items, three times each, for a total of 90 measurements. I refer to this length of a study as "long-form" and recommend (especially for initial and yearly studies) that the long method be used.

Shorter versions of the gage R&R study can be used for various reasons. The same methods are used, but instead of three appraisers, only two would be used, or instead of three trials, only two would be used, or instead of a group of ten items to be measured there would be five. Under obvious circumstances only (i.e., only a very basic evaluation is needed, the gaging process requires that a costly part be modified in some way, scarcity of parts, etc.) do I advise you to vary from the long method as it is not as precise.

6.2.2   How to Conduct a Gage R&R Study

Let's assume that we are evaluating the measurement system for an ABS plastic disc. The significant characteristic is part thickness and the measurement tool is a micrometer. The data will be recorded on a worksheet example like the one shown in Exhibit 20. Ten discs have been numbered with a grease pencil (1–10) on one side and turned over so that the operators cannot identify the number. The process will proceed as follows:

1. Make sure the micrometer has been calibrated.
2. Operator A will pick up and measure one disc with the micrometers. The measurement taken will be marked on the worksheet under the same number on the back of the disc. Operator A will continue to measure one at a time and place each measurement under the item number corresponding to the number on the back of the disc.
3. After the discs have all been turned back over so that the numbers do not show and are randomly mixed up, Operator B, and then C, will repeat the same steps as A.
4. Operator A will now begin the second trial, repeating the same process as described in item 2. Operator B and C will also complete the second trial.
5. All three of the operators will then complete the third and final trial in the same manner as the others.
6. The averages and ranges (as indicated on the worksheet in Exhibit 20) will be figured for each part and for each operator. Those numbers will then be applied to formulas outlined on the second worksheet example (see Exhibit 21 for the formulas).

6.2.3   How to Evaluate the Results

"Repeatability" is the ability of the *equipment* to repeat a reading on a part. "Reproducibility" is the ability of the *appraiser* to reproduce the gaging method with the gage and part. The equipment error and the appraiser error are combined to produce the overall gage R&R.

The individual ranges of the appraisers can be evaluated by comparing that data within the UCL (upper control limit) and LCL (lower control limit) which is the grand average of ranges multiplied by the constants D4 and D3, respectively. If any appraiser's ranges are out of control, then they could need to be retrained on the method of gaging.

Gage R&R variation can be evaluated in regard to both the process and the blueprint tolerance. When evaluating the percent of process variation, you would divide the part variation by the total variation. The percent should be as close to 100 as possible. It is not intended for the equipment and appraiser variation percent, if added to the part variation percent, to equal 100%. When evaluating variation based on the percent of the tolerance, the tolerance would simply replace the total variation factor in the calculations (as indicated in Exhibit 21). I recommend that both approaches be taken. If the R&R is over 10–30%, look at the individual percents of the equipment and appraiser to determine which, if not both, is the cause so that corrective action/s is directed at the source of the variation.

Acceptance of the amount of variation should be taken in all approaches as follows:

- Under 10% R&R error—acceptable,
- 10–30% R&R error—may be acceptable based on importance of application, cost of gage, cost of repairs, etc., and
- Over 30% R&R error—gage system needs to be improved or corrected.

If for some reason, a gage is to be accepted with more than 30% R&R, the acceptance of the gage should be put into writing by the customer or company representative who approved it. That document should be placed in the job file for future reference.

6.2.4   Helpful Hints

Always include a space for the Gage R&R to be recorded on the front of all variable SPC reports. A log sheet of all Gage R&Rs classified by part number and gage can be used for quick reference. Knowing the Gage R&R when evaluating SPC and Cp/Cpk establishes the confidence level of the data being reviewed.

6.3   Calibration

Controlling the accuracy of measuring and test equipment and measurement "masters" (check blocks and fixtures) maintenance is necessary to assure that products delivered to the customer conform to specified requirements. The calibration control system should definitely apply to all departments and individuals who use measuring test equipment (including employee-owned equipment). Quality usually has the prime responsibility for operation of the calibration system.

6.3.1   Frequency

All measuring and test equipment and devices used to determine an item's conformance to specified requirements should be calibrated. Calibration should occur upon receipt (or before initial release for use) and at regular intervals of at least one year or sooner if deemed necessary. Many devices, depending on frequency of use, will need to be scheduled for calibration at shorter frequencies. For example, if a particular pair of calipers were used daily around the clock, they could be calibrated at least once per month. If a set of micrometers is used only once per month, they would be calibrated every six months to one year.

6.3.2   Standards

All measuring and test equipment devices should be calibrated to working measurement standards, which, in turn, should have been calibrated to reference measurement "masters" that are calibrated and certified by the

National Institute of Standards Traceability (the NIST). All gages should be accurate to +/- .0005 of an inch before they are approved by Quality.

6.3.3   Records and Calibration Identification

Records should be maintained that uniquely identify each item of measuring, test equipment, and each measurement "master" with a gage identification number and name. The date of each instance of calibration, the actual measurements, and any adjustments should be recorded.

Each measuring device should be marked or labeled to show the date of the most recent calibration, the calibration performer's initials, and the recall or due date of the next calibration. Departments and individuals that use measuring and test equipment should be responsible for monitoring calibration due dates and submit the devices on schedule.

6.3.4   Gage Control

Gages used should always be maintained in a manner that assures continuing accuracy and identification that indicates the latest engineering change level (if applicable). Inspectors and test technicians should not accept measurement values obtained on measuring and test equipment that have exceeded calibration due dates.

Measuring and test equipment that has been dropped or otherwise damaged in any way, should be returned to the Quality Department and tagged as "Do Not Use." The item should be stored pending repair and calibration by the authorized testing personnel. Equipment found to be out of calibration or not within the date of periodic calibration, should be tagged "Do Not Use" and returned to Quality to be calibrated by the assigned/ certified personnel.

6.3.5   Suspect Stock

At any time a fixture or any measuring equipment is unable to be calibrated or is found to be out of calibration to the point that bogus data could exist as a result, all stock produced should be reinspected to meet inspection criteria. If parts were shipped that potentially deviated from the blueprint because of poor data collected, the customer should be notified by Quality Management immediately.

6.3.6   Helpful Hint

There are many inexpensive computer programs available for gage control. I recommend a computerized program that houses all certification information, calibration records, and Gage R&R records. Due dates can then be maintained on the computer which upon command will recognize any fixtures and/ or equipment that is due for calibration and R&R.

6.4   Gage and Fixture Maintenance

The maintenance of gages and fixtures is too often considered low priority. Unless there is a formal maintenance procedure built into the Quality System for gages and fixtures, they will inevitably be neglected. That is, until a job is waiting for first sample approval and it becomes a crisis that the tip on the indicator is bent, or the feeler gage is missing, etc. Or maybe stock is rejected and ground up because the gage indicated that it was unacceptable when, in fact, the Gage R&R "sky-rocketed" due to dirt build-up in a spring that helps locate and lock the product. In a shop atmosphere, a holding fixture can quickly gather enough dust to change a net surface by 0.01 mm or more.

   When establishing a formal procedure, don't forget to assign personnel who will be accountable for each task involved in the process.

6.4.1   The Process

Of course, there can be many variations to the process of gage and fixture maintenance that I will describe here. My intention is only to provide some basic guidelines to be considered when developing a procedure.

   Preventative maintenance should be the first area investigated in this matter. It's difficult to totally prevent the need for maintenance of gages and fixtures, but efforts in this area will indeed reduce the need for maintenance; thus, reduce the frequency of otherwise planned maintenance. The following examples could be considered preventative and reduce scheduled maintenance:

- Buy fitted covers to be placed over all gages and fixtures when they are not in use.
- Invest in retractable fixture feelers and check pins. Those that are attached to fixtures with a chain or wire will need repair and replacement more often.
- Provide adequate, safe, and clean storage areas for gages and fixtures.
- Invest in hydraulic carts to retrieve and move heavy gages and fixtures. This will provide less chance of damage to them and will reduce the likelihood of employees becoming injured by them.

Scheduled maintenance of fixtures and gages should be determined based on need and should be evaluated and updated when needed and yearly. Some items that might appear on a fixed maintenance schedule are as follows:

- Clean with mild soap and water quarterly or as needed.
- Oil all clamp joints and wipe excess oil monthly or as needed.
- Make sure all instruction plates are on the gage/fixture and up to the current Engineering Change Level each use. Report in writing to the Quality Manager if there is a discrepancy.
- Check for all pins and feelers to be intact each use. Report in writing to the Quality Manager if there is a discrepancy.
- Blow off the gage/fixture with an air hose before and after each use.
- Report any damage or missing parts immediately to the Quality Manager.

6.4.2   Helpful Hints

If possible, have a segregated area for all gages and fixtures. Develop a log sheet whereby the Inspector or production personnel must sign gages/fixtures in and out. The log sheet could also contain a checklist for the scheduled maintenance.

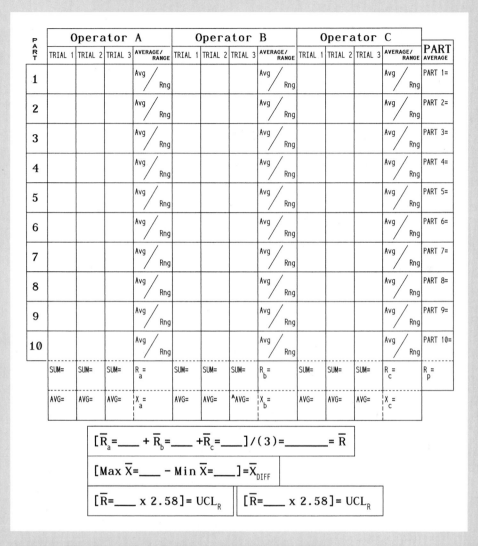

P A R T	Operator A				Operator B				Operator C				PART AVERAGE
	TRIAL 1	TRIAL 2	TRIAL 3	AVERAGE/ RANGE	TRIAL 1	TRIAL 2	TRIAL 3	AVERAGE/ RANGE	TRIAL 1	TRIAL 2	TRIAL 3	AVERAGE/ RANGE	
1				Avg / Rng				Avg / Rng				Avg / Rng	PART 1=
2				Avg / Rng				Avg / Rng				Avg / Rng	PART 2=
3				Avg / Rng				Avg / Rng				Avg / Rng	PART 3=
4				Avg / Rng				Avg / Rng				Avg / Rng	PART 4=
5				Avg / Rng				Avg / Rng				Avg / Rng	PART 5=
6				Avg / Rng				Avg / Rng				Avg / Rng	PART 6=
7				Avg / Rng				Avg / Rng				Avg / Rng	PART 7=
8				Avg / Rng				Avg / Rng				Avg / Rng	PART 8=
9				Avg / Rng				Avg / Rng				Avg / Rng	PART 9=
10				Avg / Rng				Avg / Rng				Avg / Rng	PART 10=
	SUM=	SUM=	SUM=	R_a =	SUM=	SUM=	SUM=	R_b =	SUM=	SUM=	SUM=	R_c =	R_p =
	AVG=	AVG=	AVG=	$\bar{X}_a$ =	AVG=	AVG=	^{A}AVG=	$\bar{X}_b$ =	AVG=	AVG=	AVG=	$\bar{X}_c$ =	

$$[\bar{R}_a=\underline{\quad} + \bar{R}_b=\underline{\quad} + \bar{R}_c=\underline{\quad}]/(3)=\underline{\qquad}=\bar{R}$$

$$[\text{Max } \bar{X}=\underline{\quad} - \text{Min } \bar{X}=\underline{\quad}]=\bar{X}_{DIFF}$$

$$[\bar{R}=\underline{\quad} \times 2.58]=UCL_R \qquad [\bar{R}=\underline{\quad} \times 2.58]=UCL_R$$

Exhibit 20: Gage R & R study (data worksheet)

FROM DATA WORKSHEET: $\bar{R} =$ _____ , $\bar{X}_{DIFF} =$ _____ , $R_p =$ _____

MEASUREMENT UNIT ANALYSIS	% TOLERANCE ANALYSIS
REPEATABILITY-EQUIPMENT VARIATION (EV) EV = $\bar{R}$ × 3.05 EV = _____	% EV = 100 [EV/TOLERANCE] % EV = _____ - % AV = 100 [AV/TOLERANCE]
REPRODUCIBILITY-APPRAISER VARIATION (AV) AV = $\sqrt{[(\bar{X}_{DIFF} \times 2.70)^2 - (EV^2/90)]}$ AV = _____	% AV = _____ %R&R=100 X [(R&R)/TOLERANCE] %R&R = _____
REPEATABILITY & REPRODUCIBILITY (R&R) R&R = $\sqrt{(EV^2 + AV^2)}$ R&R = _____	% PROCESS VARIATION % EV = 100 [EV/TV] % EV = _____
PART VARIATION (PV) PV = R_p × 1.62 PV = _____	% AV = 100 [AV/TV] % AV = _____ - %R&R = 100 X [(R&R)/TV]
TOTAL VARIATION (TV) TV = $\sqrt{(R\&R^2 + PV^2)}$ TV = _____	%R&R = _____ - % PV = 100 [PV/TV] % PV = _____

* IF THERE IS A NEGATIVE VALUE UNDER THE SQUARE ROOT SIGN FOR AV, THE APPRAISER VARIATION DEFAULTS TO 0 (ZERO).
* ALL CALCULATIONS ARE BASED UPON PREDICTING 99% OF THE AREA UNDER A NORMAL CURVE.

Exhibit 21: Gage R & R study (report worksheet)

7 Statistical Process Control

I will not describe the basics of Statistical Process Control (SPC), or how to calculate averages, ranges, percents, sigma, etc., in this chapter. There is already a more than sufficient amount of books available that teach the technical methods involved in charting and analysis. Throughout my career, the comment I've heard most often has been, "We know why and how to chart and understand how to analyze SPC charts; what we can't figure out is *how* to successfully implement the system and make it work." That is exactly what this chapter will focus on. As well, I will point out some of the "basics" that are often overlooked or misunderstood.

7.1 Training and Implementation

The key to successfully implementing a system or process is proper training. In regard to SPC, after the system is developed, the highest Quality and highest Manufacturing/Operations Officials should develop the training requirements.

7.1.1 Training Requirements

If your company has developed an SPC system, it would be assumed that there is at least one SPC "expert" in your company. If not, someone proficient in SPC should be hired or an individual sent to a local university or college for extensive training to become your needed expert. The system that your company develops should indeed have had input from someone with expertise in SPC.

Now, assuming that you have identified who the skilled SPC person(s) is (are) in your company, you should assess the abilities of those persons to train others. There is a very distinct difference between "knowing" a subject and "teaching" it to others. If your expert is not capable of effectively training others, one or a combination of the following approaches could be taken: either provide training to the expert on how to teach effectively or work with your local university or college in setting up a customized training program through them. For example, while your skilled SPC person is learning how to train, the Manufacturing and Quality employees could attend university or college taught classes on basic SPC (many universities and colleges will teach at your location if there is a large group requiring the training). The expert, once trained in educating others, can explain the company's need for SPC and train employees on your company's specific procedures and process.

7.1.2   The Training Approach

It is best for the trainer to know up front that the majority of the employees being taught SPC may *not* be receptive. I state this from experience in finding that overall, the majority of the population is not receptive to change— especially forced change. I have found that it is best for the trainer to inform the employees (diplomatically) that he or she is aware of the fact that many of them are not receptive to taking on an additional task. The introduction to the training could be similar to the following:

> "Today I will be instructing all of you on how to use a tool that will help the company gain control over its processes. Controlled processes *will* affect you by reducing defects and being more predictable. I am not oblivious to the fact that many of you sitting here right now will not want to be receptive to what I'm trying to teach you. It involves added responsibility to your current job and it may be hard to be happy about that. I need you all to trust me when I say that in a couple of months you will see that your involvement in using this tool requires very little time and should not make your job more difficult. The tool I am referring to is called statistical process control. It simply means that we will control our processes with a statistical technique, or with measurements. We'll call the tool "SPC" for short. Again, please keep an open mind and decide now that you can be interested in this and the training will be much easier for you to understand."

It is also very important for the trainer to realize that not everyone is as intelligent as he or she is. Many of the employees to be trained will have retained very little of the basic math they learned in grades 1–12 in school. The last thing the trainer wants to do is to overwhelm the employees beyond comprehension. Don't forget to spend as much time as needed on the following components of SPC charting:

- Adding positive and negative numbers,
- figuring range (use analogies),
- reading and understanding the scale on the graph,
- plotting a point by understanding the scale's increments on the graph, and
- figuring percents (make up "cheat-sheets" for Np-charts that have constant sample sizes).

I suggest also that you allow your employees to "get used to" SPC before expecting the process to be properly followed. This approach will take pressure off of the employees and help them be "open" to training if this is announced before training begins. For example, after stating that you are aware that they most likely are not going to be receptive, you could announce the following:

> "I am not asking that each of you leave here today and do what I'll ask of you. I will request though, that you at least *try*. Keep an open mind, *try* to understand what I'm showing you, and then just *try* to do it. I don't care if you add numbers wrong or plot a point wrong or misinterpret an out-of-control condition. I do care, however, if you don't even attempt to practice what I'll show you here today. I am not asking for perfection, I am asking for an attempt at being sufficient."

This approach will automatically cause most employees to become curious and more open-minded. A month or two later, refresher classes can be held and more emphasis put on "doing it right." By then, the initial defensive attitudes of the employees will have dwindled.

7.1.3   Implementing the SPC Program

Do not begin an SPC program until procedures have been written and the entire process is documented. The quickest way for an SPC program to fall by the wayside is for top management to not fully support it, follow up on it, and most of all, to react to it.

It is essential that tracking logs be developed to maintain records of actions being taken to increase Cp/Cpk's and reduce P-bars (average percent defective). Post history sheets at operations that depict P-bars going down and Cpk's going up. Post all action plans and any other documentation that will assure all employees that SPC is taken seriously by management and reacted to. The employees must be aware that their efforts were not futile. When management "forgets" about procedures and processes once deemed important, it is inevitable that the employees will follow suit. I recommend that the SPC process portion for which supervision and management is responsible for be relayed to the employees during training. Make sure that they know the process goes beyond their particular responsibilities and "all eyes" are relying on the data they gather.

Be patient. Expect continuous improvement in regard to the effectiveness of your company's SPC program, but do not rely on the procedures to be followed precisely in the beginning. A lot of effort should be focused on developing realistic and rational expectations for progress. I suggest that an overall plant "SPC Improvement Plan" be developed and documented as a guideline for Supervision and Management to gauge what is considered reasonable progress. This plan will help alleviate unnecessary frustration and disappointment for everyone.

After six months to one year, depending on the size of your company, the program will have become second-nature and begin to provide more accurate information. This is not to say all data before this time frame will be inaccurate, or that all data after will be accurate. As time progresses, understanding of SPC and your company's specific SPC procedures will increase. The overall system will be optimized with gained experience, thus producing more accurate information and recognized process improvements where action is taken. There will most likely always be mistakes made as a result of human error. Mistakes should be analyzed so that potentially preventative measures can be taken to eliminate or reduce the opportunity for them to reoccur.

7.1.4   Helpful Hints

If your company is implementing a new SPC program, it could be helpful to begin "small-scale" as a prototype. For instance, if your company consists of twenty blow molding processes, you may begin implementation of the new procedures on only one or two. This will allow you to monitor the new program in its real intended environment and optimize or revise it prior to full implementation.

Refresher training classes in SPC should be conducted quarterly or annually, depending on your company's stage in SPC. To make the training more effective, select a random group of employees (who have already been trained) to meet with the trainer to give him/her insight. They may reveal areas that previously were not covered, were insufficiently presented, or difficult to understand. This can allow the trainer to optimize the instruction process by better understanding the employees' training needs.

Make sure there is always someone available to the employees to give direction and answer questions in regard to SPC on all shifts. Make it *known* to the employees that those advisors are available. This will eliminate the opportunity for excuses when procedures are not followed or followed incorrectly.

7.2   Capability Studies

Once a process is in statistical control, it is assumed to be predictable. If the process has achieved a state predictability, studies can be conducted that determine the capability of it to produce a product that will meet required specifications.

7.2.1   Cp AND Cpk Indexes

Cp is the capability index that is interpreted as the tolerance width divided by the process capability, with no regard to where the process is located. The Cp index allows you to evaluate process performance by looking at process variation only.

Cpk is the capability index that considers the location and spread of the process in relation to a specified nominal requirement and tolerance. The Cpk permits you to measure the capability of a process to produce products that are within engineering requirements. For more details on Cp/Cpk analysis, I recommend *Statistical Quality Control* (Grant and Levanworth) and *Guide to Quality Control* (Ishikawa).

7.2.2   Process Potential

The first step in determining the capability of a process is to assess the *potential* of that process during the very first sample tryout. This assessment is referred to as the *process potential study* or *short-term capability study*. It serves two purposes:

1. It allows Quality, Manufacturing, and Engineering to determine whether or not the process and tool can be deemed acceptable for regular production runs, and
2. it establishes initial expectations of the process and tool to be used as preliminary guidelines.

The initial process potential study should consist of at least 50 measurements (100 are better) and placed into subgroup sizes of 5. Each cavity (if it is a multicavity tool) should be studied separately. When sampling a production ready tool and process, label or mark 50 samples or more by numbering them consecutively. It is important to keep the production in order to determine exactly how the process is behaving and to determine the cause/s of any out-of-control conditions. To eliminate as much "noise" (variation) as possible, the material used to produce samples should be of the same lot and properties. Also, the process should be stabilized so that process parameters do not have to be manipulated during the collection of samples. Use only one inspector to measure the samples collected to reduce appraiser variation. The deliberate elimination of noise during this phase of determining process capability allows us to focus on the tool and the process. It also provides us with a preliminary assessment of the process's ability to produce product at the specified nominal that does not vary outside of specified tolerances.

The process potential results can become a goal for long-term capability. The results serve as evidence that if variation of the raw material, process, tool, operator/appraiser, etc. is reduced, the same results could be achievable during production.

7.2.3   Long-Term Process Capability

The long-term process capability study typically begins with the first production run of a new product. However, remember that process capability should not be calculated if the process is not in statistical control (normal distribution). This is important because the Cp and Cpk indexes are calculated with the assumption that the process is predictable (in statistical control). If product volume requirements result in short production runs, the job may need to run a few times for enough data to be collected to determine process stability. At least 100 pieces of data (i.e., 20 subgroups of 5) should be collected before recalculating SPC control limits and determining process capability.

If production runs are long enough or data collection is frequent enough that control limits and capability can be calculated for each run, all new data should be combined with previous production data. It can be deceiving to assess each run independently. For example, if run A used ½ of the allotted tolerance and was off nominal to the high side, and run "B" also used only ½ of the tolerance but was off nominal to the low end, both might have acceptable Cpk's *independently*. If this is the case, the product should be approved, but the process consistency should be questioned. If both runs A and B were combined, the Cpk would indicate more variation, and the process would likely not be capable. There should be consistency from run to run. If there is not, then there is problem with consistency of the resin, machine, parameter set-up, auxiliary equipment, etc., that should be ad dressed. If the process has purposely been changed (i.e., different material, changed tool/s, manpower increase, plant now air-conditioned, different machine, process set-up change, etc.) and is affecting the location or variation of data being collected, then brand-new control limits and Cp/Cpk's should be calculated.

7.2.4   Tracking

The Cp and Cpk indexes should be tracked by product number for each run or for each time they are calculated. This "history sheet" could then be posted at the machine during the current run for reference. The tracking log should list the product number and characteristic being monitored at the top. The following information for each run or calculation period should include the machine number (this allows for quick assessment of which machine is more capable) and the Cp and Cpk. It is also very effective to plot the Cp and Cpk values on a chart for each calculation period. The chart could serve as the tracking log and as a visual aid for operators to see that indeed the product is improving.

Further tracking of problem solving, actions, and progress should continue to occur. I suggest the "closed-loop approach."

7.3   Closed-Loop Approach

To gain the desired benefits from an SPC program, it should be reacted to at all levels in your organization by all contributory functions. To ignore out-of-control conditions on the shop floor, or for upper management not to take action with necessary tool modifications to improve unacceptable Cpk's, is to make all other efforts in regard to SPC futile. To insure plantwide involvement in rendering an effective SPC program, it is crucial that all needed functions are involved in establishing a process that will accomplish your company's expectations. Once that process is agreed upon, it should be documented as a procedure with all responsibilities outlined.

7.3.1   The Beginning of the Process

The process outline should begin with the beginning:

- Who is responsible for putting the chart on the shop floor?
- Where do they obtain control limits to put onto the chart?
- Exactly how are control limits calculated and from what data?
- How many subgroups or individual data must be obtained to formulate limits?
- Are they simply based off of the last run (continuous process optimization program) or are all previous runs compiled (acceptable and stable process)?
- What if it is a "first run" and there are no control limits established yet?

All of these questions should be addressed by your SPC process development team and put into writing.

The procedure should state exactly how data is to be collected and transferred to the chart. It should specify the responsible individuals and define resources for them to refer to for instructions regarding frequencies, sample sizes, gaging methods, measuring instruments, etc. It is a good idea to include some of the people who will be involved in the data collection to aid in developing the SPC process. Their contribution will help to assure that all potential questions are answered and included in the procedure.

7.3.2   Different Phases of Reaction

There are three essential phases in reacting to SPC if continuous improvement of quality is to be realized. They are:

1. The floor,
2. key management, and
3. top management.

If phase 3 is disciplined and committed to quality improvement through SPC, phases 1 and 2 of "reaction" will be more effective over time.

7.3.2.1   Phase 1—The Floor

"The Floor" involves the timely response to out-of-control conditions on SPC charts as they occur. This is typically the responsibility of Manufacturing Operators, Inspectors, Foremen, and Process Engineers. All of these people should have SPC training and be updated with additional training periodically. The training should enable these employees to identify out-of-control conditions, analyze them, and constructively react to them with corrective and preventative measures.

Standards should be set in the plant that require certain functions at given frequencies to confirm Phase 1's effectiveness. For instance, it might be the responsibility of the Leader to daily check the charts on all jobs that he/she oversees. This audit could serve many purposes:

1. To verify that data is being collected and properly recorded,
2. to verify that out-of-control conditions have been identified and assignable causes and corrective actions are documented,
3. to provide a consistency to maintenance of the charts (i.e., new limits to be calculated, chart is full, etc.),
4. to be regularly available to provide direction and answer any questions in regard to SPC or the specific chart,
5. to assure the Operators through monitoring that, indeed, the charts are important (Operators are less likely to ignore the chart and their SPC responsibilities if they know their supervisor is checking them daily.), and
6. to empower Supervision to be accountable for the Floor's prompt collection of and response to the data.

7.3.2.2 Phase 2—Key Management

"Key Management" pertains to the reaction of the Manager in each department that can provide their expertise and authority in improving processes.

At the end of a run or SPC collection period, the charts should be immediately compiled into a packet with a cover report. The packet contents would consist of the actual charts, any correlating inspection and process performance records, and print-outs of histograms, pareto charts, etc. The cover report might contain at least the following information:

- Product name and number,
- type of data and chart (i.e., attribute/P-chart),
- machine number,
- run dates, and
- the latest control limits and averages.

If the report cover is for *variable* charts, it should also include the:

- Characteristic,
- blueprint specification,
- Cp and Cpk,
- latest gage R&R %, and
- sample size and frequency.

If the report cover is for *attribute* charts, it should include the:

- Total % defective,
- total sample inspected,
- sample size (if consistent),
- most common defect(s), and
- dollars not realized as a result of the amount defective.

All of this information for the cover report should fit at the very top of the report, leaving the rest of the sheet for Key Management's correspondence (refer to Exhibits 22 and 23).

Whoever is responsible for compiling the SPC packets (usually a Quality or Manufacturing individual who is considered the SPC expert), would begin Phase 2's portion of the "loop." This person would analyze the contents of the packet and document the analysis in the space provided on the cover report.

They would then pass the entire packet to the Manufacturing Manager for root cause analysis and corrective action to be recorded. That manager would then hand the packet over to the Engineering Manager, who would review the packet contents and add his/her input in regard to root cause and corrective action. He/she would then pass the packet to the Quality Manager, who would do the same before handing it in to the General Manager, whose involvement is in Phase 3.

Phase 2 enables and provokes key management personnel to respond to SPC after a run or data collection period. There is no meeting to dominate their time and they are not subjected to sit through a formal review of every chart. They give input only if they have any and there is not time taken from others when one particular manager can correct or improve the problem.

7.3.2.3   Phase 3—Top Management

Top Management is where we close the loop. The General Manager would review the cover reports simply to determine whether or not key Management appears confident in what the cause and corrective/improvement action/s will be. The General Manager should be sure that someone has taken responsibility for the corrective or improvement action. As the General Manager is reviewing the cover reports, he or she may separate them into two piles: one that will be forwarded back to the SPC expert for tracking information to be recorded, and one that warrants review in a weekly SPC or Quality meeting because cause/correction/ responsibility had not been determined.

During a review of "opened" reports, the General Manager would complete the cover report. All reports would then go back to the SPC expert so that the analysis, action/s, and responsibility can be logged onto an SPC tracking form according to its product or process number.

It should be evident why Phase 3 is the most important. If the system fails in Phase 1 or 2, it is the General Manager who assumes the last responsibility. No matter how dedicated and assertive an hourly Inspector is, they are no match to the General Manager when it comes to enforcing disciplines with Managers and Supervisors. I believe it is a fact that *if your SPC program fails, it is because Top Management did not support the system.*

7.3.3 Standards for Report Circulation and Continuous Improvement

Depending on the amount of SPC report packets that your company would normally circulate, it may be of importance to send only those reports that do not meet a preset standard. Circulating all acceptable charts could overwhelm management and take time away that would have been better spent on the priority (unacceptable) charts. However, it is also important to continuously improve even acceptable charts to optimize manufacturing processes. For this reason, I recommend that standards be set for both the circulation of information and for the quantity that is to be circulated. This method would allow the standard to continuously improve as more charts met the standard.

Let's assume that, on the average, a company generates 50 SPC reports a week (25 variable and 25 attribute). As a result of investigation, it has been determined that key management can effectively analyze and react to 20 total reports per week. Now it is concluded that this company will only circulate 40% of all completed reports. Based on recent history, if 40% of all reports were circulated, key management would see the following reports:

Variable	Attribute
Cpk's less than 1.4	P-bars above 2.5%
Gage R&R's over 24%	Value unrealized above $300 (scrap cost)

These can become the first standards for SPC report circulation.

In the meantime, the company's SPC person would maintain a chart that depicts the percent of charts circulated on a weekly basis. As SPC improvement efforts are evident, the percent of circulated charts will decrease. More charts would be filed as a result of being above the report circulation standards. So, to again reach the 40% circulation mark, the standards would be more restrictive, thus improved. Perhaps the new standards to maintain a 40% circulation rate would be:

Variable	Attribute
Cpk's less than 1.6	P-bars above 2.0%
Gage R&R's over 20%	Value unrealized above $275 (scrap cost)

As you can see, this method not only controls management's priorities but builds continuous improvement into the system. NOTE: Any chart, regardless of whether or not it meets company standards, should be circulated if it is out-of-control with no assignable causes and/or actions. Standards should *never* be lowered as a result of an increase in the percent of charts circulated. An increase in percent circulated should result in a serious investigation and corrective actions by top management.

7.3.4   SPC as Part of Total Quality Management

Many companies have a monthly Total Quality Management meeting. Some companies call this meeting "QOS" (Quality Operating System), "ITP" (Improvement Tracking Process), etc. It is usually a large multi-disciplined meeting where all functions within the company review key measurement indicators of improvement with visuals such as bar, pareto, line graphs, etc.

For SPC, the following indicators could be depicted on a graph that shows the measurable every month and what the current year's goal is:

- Average of plant Cp's and Cpk's,
- percent of circulated SPC reports,
- percent of out-of-control charts,
- top 5 priority variable charts by part number,
- top 5 priority attribute charts by part number, and
- breakdown of monthly defects on a Painter's chart.

There are many other categories and ways to measure improvement for SPC which leaves room for a company to be creative.

In this meeting, corrective actions would be determined by a team whenever goals are not being met. It is also a good way to share the company's success of SPC with other department personnel who otherwise would have no involvement with it at all.

7.4   Automation in SPC

By "automation," I am referring to any product that can relieve or take the place of manual input, calculations, reporting, etc. The most commonly used form of "automation" is an SPC software program run by computer. In this section, I will give an overview on some of the most popular types of products that you might seek, as well as some constructive advice.

7.4.1   Computers

I think the main thing to be considered in respect to purchasing a computer is your company's software needs. In commoner's terms, how much memory (random access memory) do you need and how fast (megahertz and type of microprocessor will affect speed) do you want it to perform? Will the computer be compatible to the brand of software you want to use? I suggest that you confer with at least three computer hardware sources' customer service representatives and your software salesperson before making a decision when purchasing a computer and printer.

7.4.2   SPC Software Programs

There are many SPC computer programs on the market. Expect to spend at least $2,000.00 for a good one. The inexpensive programs offer less options and have been known to have some "glitches" in some of the calculating functions. While you won't want to pay for features that your company will never use, make sure you purchase a program that is expandable in case you desire certain features in the future. Again, it is wise to call in at least three sources to explain and demonstrate the programs' capabilities (they can also lend you various "demos" for review).

A good way to get an idea of what programs your company should research is to call a few of your major customers. They can offer good advice on what to look for and inform you of what program/s they use and/or recommend.

7.4.3   Automatic Data Collectors

Data collectors can be a major asset in saving time and eliminating a large portion of human error in the recording and calculating of data. They can also enhance the availability of data by making it immediately available to any modem that is "hooked up" to retrieve it. Some even provide visual and/or audio alarms to let an operator know when there are out-of-control or out-of-spec conditions or if "unreasonable" data has been entered.

There are also many electronic measurement instruments on the market that "hook up" to and directly link data to the data collectors. These instruments (calipers, micrometers, probes, etc.) will send data to the collector with the touch of a button while the measurements are being taken. These tools are very useful in reducing human error (especially when dealing with both positive and negative numbers) and in providing "real time" SPC evaluation.

In addition to measurement instruments, are process/machine parameter data collection units. These units can be fitted to each machine and retrieve data on such parameters as cycle time, fill time, peak pressure, etc. Some are so sophisticated that once upper and lower control and specification limits are established, an alarm will sound when an undesirable cycle is in progress. It is also possible to have the unit send a signal to the conveyor when it alarms, sending the conveyor into reverse and dumping the product into a "suspect" container for evaluation. This type of automatic data collection is for those manufacturers who are very serious about process optimization and control.

7.4.4   Helpful Hints

Always think in terms of the future when considering the purchase of any of the types of products I've mentioned in this section. Especially in the field of electronic equipment and software, needs change and technology increases rapidly. Many dollars can be saved over time if purchases are made with consideration for future needs. To save a few dollars today with only current needs in mind will probably result in outdated equipment that is of no value in a couple of years. For this reason, it is very important that you ask many questions about contract provisions for automatic system updates and upgrades of the original equipment should advanced or different options become available. Remember *always* to obtain at least three quotes on any major purchase.

When considering the purchase of data collectors, equipment, and programs, ask the salesperson if you can "try it out" for a few weeks. They usually have "demos" that they loan out to potential buyers. Take advantage of the loan by implementing a pilot process using the equipment. After the pilot, formally survey every employee involved in using the equipment and its information. This will provide useful feedback on what benefits the company would derive if it was purchased.

7.5   Progressive SPC—The Elimination of "Product" Monitoring

Currently, most production organizations detect process variation by measuring its product after the fact. It is the *process* that needs to be controlled and proven capable to assure production of an acceptable product. To truly conduct a statistical process control program would mean that the actual process (its parameters) were being measured and evaluated for process optimization and stability. If control of the *process* is achieved, the application of SPC on a *product* would simply be unnecessary.

Though this concept sounds rational, a potential obstacle in implementing such a change to the norm is the customer. Many customers now require that SPC be performed on various product characteristics. In many cases, the customer will denote a number of characteristics on a blueprint or instruct you in some other way (control plans, gage designs, etc.) as to product characteristics that they want monitored with SPC. If your company were to discontinue monitoring the product and begin monitoring the process, this would disrupt the customer's established policies of expectations of the supplier.

7.5.1   How to Convince the Customer

To be successful in convincing your customer that controlling your process would better benefit them, I suggest the following:

1. Develop a procedure that will assure the customer that the transition from product monitoring to process control will be done using strict guidelines and a verification process.
2. Ask your customer to visit and review all of your current SPC history and the overall results achieved so far. This action will let the customer know that your company is very serious about consistent product quality.
3. Explain to the customer, how important it is to identify the machine parameters that will be the most indicative of changes to the product characteristics that are significant to them. Measuring a characteristic on the product is *after the fact.* The product is already finished and may have been in production like that since the last sample measurements were taken. If out-of-control product is made, there is potential of it getting to the customer. However, with

close monitoring of the process parameters that affect each particular product characteristic, you will be able to prevent inconsistent product quality by reacting to trends before the variation is measurable in the product.

4. Use my favorite analogy to present the idea: Would you put three or four different alarm systems on your house? Would you put alarms only on the few most important rooms in your house? SPC should be thought of as an alarm system—nothing is stolen yet, but it could be. The most effective alarm system would be only one (because of cost) that would "sound off" if *ANY* unwanted intrusion was made. By use of Design of Experiments (DOE), one could determine what machine parameter/s will be most sensitive to a change in the process that would eventually affect the dimensional integrity of the entire product, thus covering those product characteristics most important to the customer.

    NOTE: It is possible, with current technology, to monitor many parameters at once with the use of integral overlays of the process. The integral pictures of a process can be used to measure variation in the overall process. This type of monitoring could eliminate the need for preliminary studies, because you could automatically switch from product monitoring to integral. If a machine's entire process is repeatable, there would be no need to measure one particular parameter.

5. Show the customer case histories of certain jobs where you have switched from product monitoring to process control using statistics. Include all of the data and experiments that describe why you chose the process parameter(s) that you did. Show them examples of how trends and out-of-control conditions were reacted to in order to prevent dimensional changes to the product.

7.5.2   The Process of Eliminating Statistical Product Control

The process should be developed by a team and documented as a procedure. An ideal process would be to determine statistical process control characteristics for each new product during the advanced quality planning stages. Characteristics could be chosen by using surrogate data or conducting designed experiments to determine what process parameter/s will affect the dimensional integrity of the most important product features.

The following is an example of a procedure that could be developed to define the process of *switching* from measuring product characteristics to process characteristics:

1. Once the SPC chart for a product characteristic is in control, Cpk can be determined.
2. Once the SPC history shows stability and capability of a minimum Cpk of 1.67, maintain those conditions for a minimum of 100 subgroups of collected data over at least 5 different runs.
3. After step 2 is achieved, internal specifications will be developed for the product characteristic(s). Internal specifications will typically have a similar nominal as the blueprint, but a tighter tolerance. The tolerance should be based off of the history of 6 sigma (upper and lower control limits).

Example

Characteristic = hole size
Blueprint specification = 3.5 +/- 0.5 mm
SPC chart stability = in control for 11 runs–230 data
Cpk = 1.67+ for last 8 runs–180 data
UCL on control chart = 3.7 mm
LCL on control chart = 3.3 mm
New internal specification = 3.5 +/- 0.3 mm

NOTE: Refer to Exhibit 24, an "Internal Specification Notification Form" example.

Record the above information, obtain Management approval, and forward to all necessary individuals/departments.

4. During this period, Quality will work with Manufacturing in setting up a DOE to determine which process parameter(s) will indicate change in the dimensional integrity of the significant product characteristic(s). Once a conclusion is made, process potential studies will be conducted for evaluation with the DOE results to determine specifications for the chosen process characteristics. These will then be measured at established frequencies and monitored with SPC charting and capability studies.
5. The internal specification for the characteristic is to be added to the Inspection Instructions and checked at established inspection intervals. In turn, the characteristic will be eliminated from the SPC

Instructions and will no longer be monitored except on a pass/fail basis during routine inspection.

6. The SPC Coordinator is to maintain a "master" log sheet of all jobs in which SPC has been discontinued and include the new internal specification and ending Cpk. The log could simply contain a copy of all "Internal Specification Notification Forms."

7. If during a routine inspection, any measurement is found to be outside of an internal specification, the Inspector should notify the SPC Coordinator and reject the stock for evaluation. Quality should then consider reimplementing the previous "alarm system" (SPC monitoring of the product characteristic) at the original blueprint specification. At the same time, a serious evaluation of the process should occur to determine why its SPC methods failed to predict a change to the product characteristic. Once corrected, monitoring of the process can resume.

7.5.3   Helpful Hints

A problem that most of us will encounter in attempting this type of process will be gaining an "in-control" condition for the duration needed to justify eliminating the product SPC. This is because there is much more to contribute to overall variation when measuring a product as compared to a process. The main thing to remember here is to be very disciplined in identifying and correcting assignable causes when a chart goes out of control. If this is done correctly, the out-of-control conditions can be eliminated from calculations when determining the stability of the product characteristic.

RUN DATES:_____TO_____ PART NUMBER:_____ DESCRIPTION:_____

CHARACTERISTIC:_____ SPECIFICATION:_____ Cp=_____ Cpk=_____

MACHINE NUMBER:_____ LAST GAGE R&R PERCENTAGE:_____ SAMPLE SIZE:_____ FREQUENCY:_____

PROCESS RUN ANALYSIS

S.P.C. COORDINATOR:_____

_____ INITIALS:_____ DATE:____

ROOT CAUSE ANALYSIS

MANUFACTURING MANAGER:_____

_____ INITIALS:_____ DATE:____

(LIST CORRECTIVE ACTION/S: MACHINE, TOOL, EQUIPMENT, ETC. – NOTE MAJOR CHANGES & THEIR EFFECTS)

ENGINEERING MANAGER:_____

_____ INITIALS:_____ DATE:____

QUALITY ASSURANCE MANAGER:_____

_____ INITIALS:_____ DATE:____

ACTION PLAN

PLANT/GENERAL MANAGER:_____

_____ INITIALS:_____ DATE:____

RECOMMENDED FOLLOW-UP ACTION:_____

_____

Exhibit 22: SPC variable report (cover sheet)

JOB NUMBER:_____ RUN DATES:_____

PART NUMBER:_____ PRESS NUMBER:_____

PART DESCRIPTION:_____ OTHER:_____

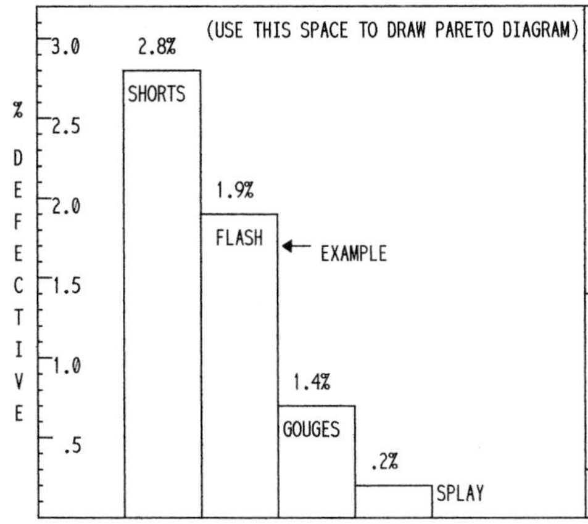

```
  ┌─3.0   2.8%   (USE THIS SPACE TO DRAW PARETO DIAGRAM)
  │      ┌─────┐
  │      │SHORTS│
% ├─2.5  │     │
D │      │     │
E │      │     │
F ├─2.0  │     │  1.9%
E │      │     ├─────┐
C ├─1.5  │     │FLASH │ ← EXAMPLE
T │      │     │     │
I │      │     │     │
V ├─1.0  │     │     │  1.4%
E │      │     │     ├─────┐
  │      │     │     │GOUGES│
  ├─.5   │     │     │     │  .2%
  │      │     │     │     ├────┐
  └──────┴─────┴─────┴─────┴────┴ SPLAY
```

TOTAL % DEFECTIVE:_____

TOTAL PIECES INSPECTED:_____

VALUE UNREALIZED:_____

MOST COMMON DEFECT/S:_____
_____

IN TOP 5 HIGHEST % DEFECTIVE: YES ☐   NO ☐

LAST RUN INFORMATION

% DEFECTIVE: _____ # INSPECTED:_____

VALUE UNREALIZED:_____ PRESS #:_____

MAJOR DEFECT:_____

BEST RUN-% DEFECTIVE:_____

PROCESS RUN ANALYSIS

S.P.C. COORDINATOR:_____
_____
_____INITIAL:_____ DATE:_____

ROOT CAUSE ANALYSIS/CORRECTIVE ACTION

MOLDING MANAGER:_____
_____INITIAL:_____ DATE:_____

ENGINEERING:_____
_____INITIAL:_____ DATE:_____

QUALITY MANAGER:_____
_____INITIAL:_____ DATE:_____

PREVENTATIVE ACTION

_____
_____EFFECTIVE DATE:_____

Exhibit 23: SPC attribute report (cover sheet)

```
PART NUMBER_____ DESCRIPTION_____

Characteristic_____

Blueprint Specification_____

Type of Variable Chart_____

Time Period 'In Control'_____

Time Period Cpk=/> 2.0 _____ Ending Cpk=_____

S.P.C. Chart UCL=_____ LCL=_____


New Internal Specification=_____
                           NOMINAL CAN VARY BUT TOLERANCE MUST BE SMALLER THAN BLUE-
                           PRINT TOLERANCE AND WITHIN BLUEPRINT SPECIFICATION.


S.P.C. Coordinator_____ Date_____

Quality Manager_____ Date_____


* FILE ORIGINAL IN PERTAINING S.P.C. FOLDER
* REVISE INSPECTION/OPERATOR INSTRUCTIONS ACCORDINGLY
```

Exhibit 24: Internal specification development (notification form)

8   Process Control and Optimization

8.1   Design of Experiments

Designed experiments are used in the advanced quality planning stage for parameter design. They can also be used later on in a program to provide insight into a process that begins to produce unacceptable product or is being further optimized. Either way, a Design of Experiment (DOE) can be an excellent tool for revealing the nature of a process, how various factors affect quality characteristics, and provide the information necessary to know exactly where investigated factors should be set.

8.1.1   Beginning an Experiment

A DOE begins with brainstorming and the use of surrogate data (if available) to identify factors that are thought to contribute to variation in the quality characteristic(s) being optimized. The quality characteristic is usually a variable, (hole sizes, strength, etc.) but can also be an aesthetic feature (visual) with a Taguchi experiment (refer to Section 8.1.2, "Different Methods of DOE"). Whichever, the characteristic must be able to be measured and the method of measurement determined. Factors are identified as influences such as: the process (particular machine parameters), material, manpower, one machine vs. another, and the environment. A "Cause and Effect" diagram (sometimes called a "Fishbone chart") is a useful tool that should be used during the brainstorming process. Refer to Exhibit 25 for a blank Cause & Effect diagram form.

8.1.2   Different Methods of DOE

Before continuing, it is important to inform you that there are many different types of experimental designs and methods. There are classical designed full and fractional factorial experiments (R.A. Fisher/Box, Hunter, and Hunter). Also Taguchi orthogonal experiments can allow one to form loss functions for responses and measure the effect of noise factors that cannot be controlled in-house. Orthogonal and fractional designs allow one to look at more factors with less trials by assuming that all two-level and higher interactions are not expected. The differences between a full factorial and a fractional factorial or Taguchi-type DOE will become apparent as I proceed.

8.1.3   Continuing the Preparation for Setting Up the Experiment

Once factors are determined, depending on the type of experiment to be used (or the type needed), levels will be chosen for each factor. For example, if "injection speed" is a factor, two levels could be chosen:

1. Lowest possible injection speed, and
2. highest possible injection speed.

A classical experiment, upon completion, will allow you to predict exactly how fast or slow injection speed should be (it could fall anywhere between the highest and lowest level). The classical full experiment will also allow you to recognize any interactions between the injection speed and the other factors included in the experiment. Note: More than two levels can be chosen in any experiment if a quadratic affect is anticipated or if you are evaluating a factor that does not have a high/low setting range. For example, if one of the factors was material type, there may be four different types of materials to be tried, resulting in four levels for that factor (material).

The use of a fractional factorial or Taguchi design would result in a smaller amount of trials to be conducted (saves money), but interactions are typically not evaluated. If they are, there ends up being more trials to conduct as in a full factorial design experiment. A Taguchi experiment is not precise in predicting where (in between the high and low levels) the "injection speed" should be set. The experiment results will simply indicate which level was best. For this reason, instead of choosing the extreme high and extreme low settings for an injection speed, it would be best to choose levels within a smaller range in the area where best performance would be expected.

8.1.4   Set-Up Example

The following is an example of the three components used to set up a DOE:

1. Quality characteristic: warpage
2. Factors:
    a. injection pack time
    b. pack pressure
    c. barrel temperature
3. Levels:
    a. 1: 1.5 seconds
        2: 3.5 seconds
    b. 1: 120 pounds
        2: 130 pounds
    c. 1: 480 pounds
        2: 500 pounds

If three factors at two levels each were designed into a classical full factorial experiment, all possible combinations would be tried while stabilizing all other machine parameters, the environment, the manpower, and material. Every combination tried would result in eight different trials to be conducted:

	a	b	c	(factors)
1	1	1	1	
2	1	1	2	
3	1	2	1	
4	1	2	2	
5	2	1	1	
6	2	1	2	
(trials) 7	2	2	1	(levels)
8	2	2	2	

Keep in mind that all possible interactions ($a \times b$, $a \times c$, $b \times c$, etc.) can be determined with this type of design. However, as you can see, this type of experiment could be very lengthy and costly if you wish to evaluate a large number of factors (if the experiment consisted of seven factors at two levels each, 128 trials would have to be conducted!).

If three factors at two levels each were designed into a fractional factorial or Taguchi experiment, a few trials would be conducted that may result in an optimum combination that was not even tried during the experiment. This type of experiment will not allow you to measure potential interactions between factors. For the three factors to be evaluated at two levels each in this manner, would result in only four trials to be conducted:

		a	b	c	(factors)
	1	1	1	1	
	2	1	2	2	
(trials)	3	2	1	2	(levels)
	4	2	2	1	

Remember that during any type of designed experiment, all other parameters must be stabilized during experimentation.

8.1.5   Orthogonal Arrays

When using the Taguchi DOE, a whole slew of orthogonal arrays are pre-made for one to choose from. For instance, the above array is referred to as an $L4(2^3)$, denoting four trials, two levels, and three factors. A common mistake made when setting up this type of experiment is to pick an array and then choose the factors to "fill it up." For example, there is no "pre-made" array to look at five factors, so many would pick the next size available array (an $L8(2^7)$ array) and end up choosing two extra factors that weren't originally planned into the experiment. To avoid this, someone should be trained on how to use Taguchi's linear graphs to aid in modifying arrays to fit your intentions, while keeping the array orthogonal.

8.1.6   Preparing to Conduct the Experiment

After the experiment is designed (set up), you will decide how many samples are to be collected for each trial. If the quality characteristic (item being measured) is a variable, at least ten samples should be collected for each trial. If the characteristic is an attribute (Taguchi experiment), at least 50 samples should be collected for each trial.

Trial and data sheets should be prepared for each trial. If trial #2 requires that factor *a* be at level 1, factor *b* at level 1, and factor *c* at level 2, the trial #2 worksheet should specify exactly what factors are to be changed to what setting. The worksheet should also contain special instructions (DO NOT CHANGE ANY OTHER PART OF THE PROCESS DURING THIS TRIAL!), and a section for the actual data to be recorded.

Once worksheets have been prepared to show the set-up of each trial and provide space to record results, they should be mixed out of order so that the trials are conducted in a random manner. To make sure that they are kept in random order, staple them (refer to Exhibit 26). If a Process Engineer, Foreman, or someone other than the preparer is conducting the experiment, attach a cover letter to the worksheets that explains how the experiment is to be conducted. It should include the length of time that each set-up should be stabilized before samples are collected, the number of samples to be collected, and what is to be measured and how.

8.1.7   Conducting the Experiment

Get the process running at what is considered the "best" set-up thus far. Record the settings of all parameters, the material lot numbers, any auxiliary equipment settings, tool watering diagram, number of operators, and whether the machine is running automatic, semi-automatic, etc. Highlight the parameters that are defined as "factors." *Only the highlighted areas are to be changed for each trial.*

Set the factors at the setting that the first trial indicates (level 1 or level 2), and let the machine cycle for at least 1 hour (a half-hour will suffice if no heat-related parameters are included in the experiment). All trials should run the same amount of time. Collect the number of samples needed at the *end* of the trial *just before* the next trial begins. Measure the samples and record the data on the worksheet. If a variable measurement is being taken, it is important that the samples from each trial are measured after the exact same amount of cooling as the other trials' samples. Consistency in all areas is very critical.

Once all trials are completed, samples measured, and data recorded, the results can be calculated.

8.1.8   Processing the Data

Processing the results of a DOE that has been conducted is complicated. I suggest software for this laborious task as well as formal education for the individual(s) who will be processing the results. The results will reveal just how much influence each factor had on the quality characteristic. They will also indicate at what level each factor should be set to produce the best achievable results.

8.1.9   Confirmation Run

Once results are achieved, a "confirmation run" should be conducted with the process set up exactly as it was stabilized during the experiment. The factors that were evaluated should be set at the level that the conducted DOE results indicate. The confirmation run should consist of about 300 cycles with at least 100 samples measured and charted (at least 200 if you are confirming an aesthetic characteristic). If a confirmation run fails, the following should be considered:

1. There could be a noise source (something that is introducing variation into the overall process) that was not controlled or measured during the experiment.
2. The factors chosen may not have included one of the major contributory influences to variation of the quality characteristic being measured. For example, a process parameter that causes significant change to the quality characteristic was not included in the experiment.
3. It could be possible that a mistake was made in recording or calculating data or in interpreting the results.
4. There could exist a quadratic effect (nonlinear) which would necessitate three or more levels being incorporated for certain or all factors that were chosen.
5. There could exist an interaction between two or more factors that was not measured during the experiment or that were misinterpreted.

8.1.10   Helpful Hints

After having a couple of individuals formally trained in DOE, have them pull "key players" together and present an overview of what DOE is about. I suggest that a one-page summary form be developed to retain the overview and results of every DOE conducted (refer to Exhibit 27 for an example). The back-up data and graphs can be filed for reference, but the one-page summary forms can be stored in a binder for quick reference on the shop floor.

Always determine cost savings for processes optimized through the use of DOE to justify the cost of the study. The savings can also be included on the summary sheet.

8.2   Process Parameter Specification Development

Resin materials that your company may purchase and use already have established standards for processing. These standards are easily obtained from the supplier and specify such windows (tolerances) as front, middle, and rear zone barrel temperatures, nozzle temperature, etc. Using material processing standards, mold information and surrogate data, the overall cycle time for new products is usually established during a quoting process before there has been the opportunity to actually run the material. It is during the first sample run, either with a production-ready or a prototype tool, that the nature of the overall process in relation to the product and tool design can be evaluated.

The sample run should be used design the parameter's windows that will produce an optimum quality product while maintaining the original estimated cycle time. Many problems will appear during a production program if the initial sample run is simply conducted to provide the customer with something to approve. If this is the goal, the cycle time may be significantly increased until "jewelry" is produced, and further down the road no one will understand why "we cannot make a good product."

8.2.1   Developing Resin Processing Resources

All available information in regard to process melt temperature of materials should be compiled into a reference list for manufacturing. This information could be contained in a manufacturing manual and be accessible to all those involved in machine set-up/processing on the floor. The list could also contain special information, for example: "Flame retardant grades of resin can be 50 °F less than specified" or "Glass/mineral fillers may require higher heats."

Resin shrink ratios, drying temperatures, and other similar information or lists could also be made available to manufacturing. In addition to providing valuable information, lists of this type should result in an increased knowledge of the characteristics of various resins by the manufacturing personnel. The more they know, the better experts they become at processing.

Making resources such as these easily obtainable by manufacturing personnel will reduce the chances of mistakes due to lack of or difficulty in obtaining information. It is important to make it clear that *these resources are only guidelines*. It should not be assumed that if the melt temperature of polyester is from 450 to 500 °F that any heats between that range will be acceptable. That is only a window out of which another window would need to be developed. It is possible that this particular tool/product design in a particular machine processes the optimum quality product at a temperature between 480 and 500 °F.

8.2.2   Taking Advantage of the Initial Sample Run

During the sample run, many studies can be conducted to aid in developing the initial process set-up. Testing viscosity and shear rates, studying rheology behavior of the resin, conducting a gate-freeze study, cavity fill-time analysis, regrind effects, actual resin melt temperature studies, etc., can educate the processor about the behavior of the material in a particular machine and mold. As each study is conducted, the process parameters can be manipulated as each study indicates optimum levels.

To wrap up this initial stage of developing parameters, a DOE can be conducted to enlighten the processor of any remaining factors that influence product quality. The DOE can also be used to determine the tolerance (window) around each significant parameter setting.

8.2.3   Documented Process Parameter Settings

Once parameter settings and tolerances are developed, they should be recorded on a "process set-up sheet" for that particular tool/product. It should be considered the "blueprint" for the process and be adhered to the next time the job is set up. And like a blueprint, it should be treated as a "living document," eternally undergoing improvements and updates. And just as the product would be rejected if it did not meet a blueprint specification, similar actions should be taken when the process runs out of specification or control.

8.3   Process Control/Verification

Just as the product is subjected to verification to standards and specifications during the process at given frequencies, so should the process set-up. Upon setting a tool and beginning a process, all parameters, water line hook-ups, auxiliary equipment, resin regrind percent, etc. should be staged to comply to the process set-up sheet's required criteria. The process specifications can be recorded as criteria at the top of a log that remains at the machine (refer to Exhibit 28 for an example). Once the process has produced an acceptable first sample, the actual parameter readings could be logged in the first row and verified to the specifications (they should fall within the tolerances).

8.3.1   The Control Benefits of Verification

After the first acceptable sample, at established frequencies (every couple of hours), the process could be reverified by the Foreman and the actual readings entered onto the log sheet. There are at least four major benefits to be derived from practicing this discipline:

1. It will serve as a record for reference. Just as the quality inspection record would be used for traceability information in the event of the product becoming or later found to be unacceptable, so would the process record.
2. It can be used to view how each parameter setting is varying from the nominal. Charts can be created from the data of chosen parameters to provide a picture of the behavior of particular parameters. It can equip you with valuable information to justify improving nominal values and tolerances.
3. It will alarm the data collector if a parameter is functioning outside of its specification limits. That would be a good indication that the

product being produced has changed and should be immediately inspected. Also, it may be indicating that there has been a significant change in the material (especially with decoupled type molding) that should be investigated.

4. It serves as a reminder to the manufacturing personnel responsible for process control that "blindly tweaking knobs" on the machine is unacceptable. If the process needs to be changed for quality purposes, manufacturing should be instructed to work within the tolerances specified on the process set-up sheet.

The process verification log should also be filled out whenever a change is purposely made to the process.

8.3.2   Deviating from the Process Specification

Inevitably, at times, certain quality concerns will arise that will necessitate a change to a parameter that will cause it (the parameter) to fall outside of the specified tolerance. If this occurs, there should be a documented deviation procedure for the manufacturing person to follow. The procedure should at least consist of: highlighting that individual entry on the verification log, filling out some sort of deviation request form, having a sample of the product inspected by Quality as if it were a first sample, and getting written approval of the process from the Quality department.

8.3.3   Reviewing the Data

At the end of every run, the process verification logs should be reviewed and evaluated for potential process optimization/ specification changes.

8.4   Continuous Process Optimization

There are many tools, disciplines, software programs, etc., that can be used to aid in continuous process optimization efforts. However, there are two very inexpensive disciplines that most companies already have in practice but do not extend the practice effectively to continuously optimize processes.

These disciplines are "Process Verification" and "Failure Mode and Effects Analysis" (FMEA).

8.4.1   Evaluating Process Verification Data for Process Optimization

The method of evaluating process verification data to optimize processes is one of the most pragmatic systems available. It is logical because one would assume that if the process continued to run, it was producing an acceptable product. This would mean that all of the data recorded was "acceptable." If the data for a certain parameter shows that the average setting during the course of the run was off target (not at nominal), then evidently the specified target is not best (at optimum). For example, if the specification for "injection pack time" is 2.5 seconds +/- 0.5 seconds, and the verification data throughout the run shows an average of 2.7 seconds, the specification might be changed to 2.7 seconds +/- 0.3 seconds. By continuously practicing this evaluation and updating parameter nominals to match the most recent average, the process is undergoing continuous improvement.

In situations where the parameter was off target due to variation in some other part of the process, the nominal should not be changed without proper root cause analysis. The root cause would need to be identified and expected to remain unchanged if the parameter nominal is to reflect the new average.

Further enhancement of this process of continuous improvement would result from keeping a separate log of all specification changes made to the process. This will enable you to recognize patterns. It certainly wouldn't make since to "flip-flop" back and forth every other run with two nominals. If, for instance, one run shows an average of 2.7 seconds for injection pack time, and the next shows it back to the original 2.5 seconds, continued behavior such as this would indicate that the nominal should be set at 2.6 seconds.

A separate record of all changes will also provide you with the information to recognize ongoing upward and downward trends. This type of information could permit you to "skip" a step. For example, the Foreman or whoever is responsible for the process has been instructed to keep the process parameters within the specified tolerance. A steady upward trend seen on "pack pressure" may be representative of the Foreman using the high end of the tolerance every run because the product quality is best with the setting at the high end. If the "pack pressure" specification has increased by increments of 1 for a few times in a row, raise it by 2 for the next run and continue doing this until the data stabilizes. As long as the tolerance remains the same, the low end is still available if the "skipping" takes you past the optimum setting.

8.5   FMEAs—Are They Really "Living Documents?"

In the Section 2.4, "Failure Mode and Effects Analysis," I have already explained how a FMEA is an excellent tool for continuous improvement of the process/product. I am including it again in this chapter to reiterate its usefulness. I also want to emphasize the fact that the whole FMEA process is only about 1% effective if this particular phase is ignored (as it often is). The highest risk priority number (RPN, refer to Section 2.4, "Failure Mode and Effects Analysis"), if consistently a focus for improvement, will inherently result in continuous improvement of the process.

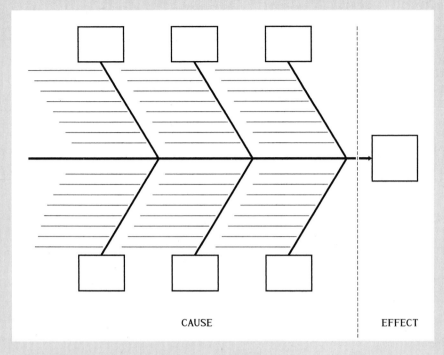

CAUSE   EFFECT

Exhibit 25: Cause and effect diagram

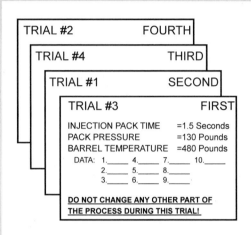

TRIAL #2   FOURTH

TRIAL #4   THIRD

TRIAL #1   SECOND

TRIAL #3   FIRST

INJECTION PACK TIME   =1.5 Seconds
PACK PRESSURE   =130 Pounds
BARREL TEMPERATURE   =480 Pounds

DATA: 1._____ 4._____ 7._____ 10._____
        2._____ 5._____ 8._____
        3._____ 6._____ 9._____

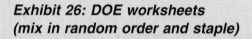

**DO NOT CHANGE ANY OTHER PART OF
THE PROCESS DURING THIS TRIAL!**

*Exhibit 26: DOE worksheets
(mix in random order and staple)*

PART NUMBER _____ DESCRIPTION _____

QUALITY CHARACTERISTIC _____ MEASUREMENT TOOL/TECHNIQUE _____

INSPECTOR _____ EXPERIMENT CONDUCTOR _____

D.O.E. DATE _____ ANALYSIS COMPLETION DATE _____ CONFIRMATION DATE _____

DATA TYPE (ATTRIBUTE/VARIABLE) _____ ARRAY TYPE _____ # OF REPETITIONS _____

IF ATTRIBUTE: # OF CLASSES & DESCRIPTION _____

FACTORS & LEVELS:

FACTORS	LEVELS	FACTORS	LEVELS
A.	1 : 2 : 3 :	G.	1 : 2 : 3 :
B.	1 : 2 : 3 :	H.	1 : 2 : 3 :
C.	1 : 2 : 3 :	I.	1 : 2 : 3 :
D.	1 : 2 : 3 :	J.	1 : 2 : 3 :
E.	1 : 2 : 3 :	K.	1 : 2 : 3 :
F.	1 : 2 : 3 :	L.	1 : 2 : 3 :

ANALYSIS

MAJOR CONRIBUTING FACTOR/S TO: THE MEAN (AVERAGE) _____ THE VARIATION (RANGE) _____

RECOMMENDED CONFIRMATION (FACTORS/LEVELS) _____ PREDICTED PROCESS AVERAGE _____

ANALYSIS COMMENTS: _____

_____

_____ SIGNATURE: _____

CONFIRMATION

PROCEDURE: _____

_____ PROCESS AVERAGE (RESULTS): _____

COMMENTS: _____

COST SAVINGS: _____

Exhibit 27: DOE summary

Exhibit 28: Process set-up and verification log (injection molding)

MOLD NUMBER: _____   PART NUMBER: _____   DESCRIPTION: _____   PRESS NUMBER: _____

PARAMETERS	HEATER ZONES				TIMERS						PRESSURES				TRANSFER			SCREW	VELOCITY	MOLD TEMPERATURES			RESIN				COMMENTS
	NOZZLE	FRONT	MIDDLE	REAR	PACK	ACTUAL FILL	HOLD	CLOSE	CYCLE	MOLD OPEN	ACTUAL PEAK	PACK	HOLD	BACK	POINT TRANS	SHOT SIZE	CUSHION SIZE	RECOV TIME	PACK SPEED	A- PLATE	B- PLATE	GALLON PER MINUTE	% REGRIND	% MOISTURE	DRYER TEMP	LOT #	MAKE NOTES FOR ANY CHANGES. CIRCLE OUT-OF-SPEC. ENTRIES AND DESCRIBE REASON/S FOR DEVIATION.
CRITICAL																											
SPEC. HIGH																											
LOW																											
DATE/INIT																											
DATE/INIT																											
DATE/INIT																											
DATE/INIT																											
DATE/INIT																											
DATE/INIT																											
DATE/INIT																											
DATE/INIT																											

Exhibit 28: Process set-up and verfication log (injection molding)

9　Problem Solving

9.1　Various Techniques of Problem Solving

Problem solving efforts in a factory should be driven in a systematic manner by established and documented guidelines. Though problem solving for many seems to come naturally as they progress logically towards solutions, efforts could be more effective if the team approach was used. Teams need guidelines, decision making empowerment, and support. Rather than reiterate throughout this chapter that problem solving is most effective when using multi-disciplined teams, I will state that this part of the guideline should to be applied to every problem solving effort. Even "experts" in particular areas can gain invaluable insight from those removed from a situation or at a level thought to be ignorant or biased about a situation.

9.1.1　Problem Solving Tools

There exists a massive quantity of problem solving tools in the industry. The following is a list to give you a brief idea of the diversity of tools available:

- Statistical Process Control (SPC),
- Capability studies (Histograms, Cp and Cpk studies),
- Pareto Analysis (and other data "visuals"),
- Cause and Effect Analysis (Fishbone charts),
- Root Cause Exhaustion (brainstorming),
- Quality Function Deployment (QFD),
- 7 and 8 Discipline Problem Solving Methods,
- Design of Experiments (DOE)
- Employee Involvement Groups (empowering floor-level personnel), and

- Procedures (unique processes developed by individual companies that provide a system to correct specific and categorical problems).

As a reference for all employees, a list should be made available that includes all of the types of problem solving tools that your company utilizes. The list can be integrated into a matrix that defines the following:

1. The best tool to use for ongoing problem prevention, trouble-shooting, advanced quality, etc.,
2. which department or individual is experienced in serving as a resource in using the tool, and
3. any specific company procedure that relates to why, when, and how the tool is to be used.

9.1.2   Disciplined Problem Solving

Every problem solving tool is either part of, enhances, or incorporates a disciplined problem solving method. I referred to 7 Discipline (7-D) and 8 Discipline (8-D) problem solving tools previously in this section. These are two very similar tools that were developed and used by automobile manufacturers. Variations of this process are called by different names and used in other industries. The point is that an effective problem solving method should contain the following elements:

- A team,
- clear understanding of the problem,
- causes and root cause of the problem,
- immediate actions and problem containment plan,
- actions that correct the problem,
- actions that prevent the problem from occurring again
- verification (preferably data) that the corrective/preventative actions were successful, and
- team approval (unanimously) to close the issue.

All of these components should have names, due dates, completion dates, details, responsible individuals, etc., associated with them. A report encompassing all of the information should be considered a living document and be updated as needed until it is considered "closed" (all components satisfied). To ensure the process is followed completely, a procedure should be written to accompany a form that provides a brief description of each element and space for the needed information to be recorded.

9.1.3   Unique Problem Solving Procedures

Every process within your company has potential for problems to occur. However, some processes (especially those intended to track and find problems) will necessitate the development of special problem solving procedures. For example, in conjunction with the "Process Verification" process, there could be guidelines that outline the steps to be taken when a product quality problem occurs. The first step might specify that the material, auxiliary equipment, etc., be verified as correct and working. The second step could then specify that a problem solving log be referred to for the specific problem. Perhaps "splay" appeared on the product—the log would prompt a moisture check of the material and indicate which parameters should be manipulated within their tolerances to correct the problem. If the problem still remained, the next step might require a Process Engineer to become involved, and maybe even a process deviation would have to occur to correct the problem. Whatever the solution, it should then be added to the problem solving log under the problem of "splay" for future reference.

What happens if an SPC point goes out of control? The problem solving methods should be outlined in a procedure. Flow charts of problem solving processes are an excellent visual aid in assuring the proper sequence of steps are followed in solving problems, (see Exhibit 29 for an example of a problem solving flow chart).

9.2   Customer Concerns

In this chapter on problem solving, special attention is being given to customer concerns for what should be apparent reasons. Unhappy customers affect both current and future business with a very negative impact that could result in the ultimate threat of lost business. The difference between a favored company and an unfavored company is not necessarily based on a number of concerns/problems realized. The difference is between those companies that react effectively and those that portray a sense of apathy to the customer by not reacting effectively. This fact lends truth to the saying that "Problems are really only opportunities in disguise."

9.2.1 Initial Reactions

A log with categories that prompt the recording of vital information at the inception of learning of a customer concern should be "at hand" for the person responsible for receiving customer complaints. Refer to Exhibit 30 for an example of a customer concern log. Whether reported by the customer via telephone, mail, fax, or in person, the log should be filled out *immediately* while the complaint is being received. The log might contain information similar to the following:

1. A concern serial number issued by your company for traceability purposes,
2. the customer's traceability number, if applicable,
3. the product number and description,
4. the amount of product in question or an estimated percent of product that is causing the problem,
5. the reason for the concern (complaint, rejection, accumulation, etc.),
6. the customer's name, location, and the contact person's name,
7. the phone and fax number to reach the customer contact,
8. the date that the concern notice was received,
9. product traceability information (date wheel information, production date, packing slip information, container label information, Quality approval stamp number, etc.), and
10. the immediate planned action: quarantine of in-house stock for auditing, scheduled visit to customer location, direction to sort or rework the questionable product, authorization for the customer to scrap or return the product, the initiation of a disciplined problem solving report, etc.

9.2.2 Plantwide Communication of Customer Concerns

Immediately after receiving the concern, the information obtained for the log should be transferred and magnified into a customer concern report that will be issued to every department and posted for all employees. This report should be written onto a pre-made form that will insure that all "bases are covered," (see Exhibit 31 for an example of a concern report used in a molding plant).

9.2.3   Corrective/Preventative Action Planning and Execution

The next step is to assign a team captain to the concern. The team captain should be the individual most able to find the root cause and take corrective action. The team captain should use anyone else necessary to solve the problem and have them as a resource during the course of the concern being "open."

The same process described as "Disciplined Problem Solving" in this chapter (Section 9.1.2) should be followed (see Exhibit 32 for an example preventative action form):

- Team,
- problem description,
- root cause,
- immediate actions,
- corrective actions,
- preventative measures,
- verification, and
- team sign-off.

9.2.4   Tracking and Follow-Up Process

Either on a separate tracking log or by adding to the initial customer concern log, concerns should be tracked as to their status. This portion of the records should maintain the following:

- Team captain,
- projected closed date, and
- actual closed date.

In the Quality Meeting that normally occurs (preferably weekly), the tracking log should be reviewed and updated. As well, every problem solving report that has been completed since the last meeting should be reviewed. A representative from each department present at the Quality Meeting should also "sign-off" on the report if they agree that the issue is justified in being considered "closed." If all required signatures are not obtained because someone feels the concern should not be closed, the team should seriously discuss leaving the concern open until further verification can be obtained.

9.2.5   Helpful Hints

Never tell a customer that the problem is not your company's fault. If, indeed, the problem is thought to be occurring in the customer's facility or with another one of their suppliers, diplomatically express this *possibility* while offering your assistance in root cause analysis. This will enable the customer to remain receptive and may even result in you becoming a "hero" in solving the problem.

Never take a stance with the customer regarding a problem that you personally have no control in correcting. You represent your company and should behave as if you yourself shipped the product that is of concern. Never bump the responsibility of solving a problem or taking action onto another individual or department when communicating with the customer. If solving the problem would require the expertise of someone else, take responsibility for getting that person or department involved *with you.*

9.3   Proactive Problem Solving

Proactive problem solving involves searching for, finding, and preventing potential problems from occurring. SPC, preventative maintenance studies, process verification, use of FMEAs, etc., are tools used in proactively solving problems. However, being "proactive" can also relate to finding and correcting problems that have already surfaced but have not yet been brought to your attention by conventional means. In many cases, the customer is responsible for making you aware of problems. In addition to prevention efforts, actions can be taken to investigate and correct problems at the customer location before the customer's process of reporting them to you actually takes place. Two excellent ways of accomplishing this are through unsolicited "visits" to the customer and frequent inquiries as to service and warranty information in regard to your company's product(s).

9.3.1   Warranty and Service Center Reviews

There should be a written procedure that describes a process whereby Sales Representatives (or another designated individual/department) are to obtain warranty and service information at an established frequency for all of the company's products. Reports could then be published to Manufacturing and Quality so that actions can be taken and verified *before* the customer contacts your company with the concern.

Let the customer know that your company is "on top of the problem" by contacting them with information regarding your findings and actions.

9.3.2   Scheduled Customer Visits

In most situations where a visit is made to the customer, it is because the customer requested or recommended it. When this is the case, there is already an established agenda of how your time will be spent upon visiting that customer. To establish good relationships with the customer and have opportunity to investigate potential and existing problems before the customer requires it, it is necessary to visit the customer when no problems are pending.

Scheduled customer visits should be planned on a yearly basis and scheduled. The plan could state which customer(s) will be visited each month and the individual responsible to conduct the visit. I suggest that individuals from both Quality and Manufacturing schedule these visits (as opposed to a Sales Representative), because they would be more apt to detect and solve any problems. Once the plan is completed, a procedure should be written to describe how the visit is to be set up and how the process of investigation is to take place.

At least two weeks prior to the intended visit, contact the customer (usually the Receiving Inspection or Supplier Quality function) and schedule a specific date and time for your visit. Explain to the customer that your intention is to view how your product is functioning in their process. Let them know that your intention is to gain any knowledge possible that will help your company improve product quality and/or service. Most customers will joyfully welcome you. However, if you get a negative response, at least the customer will know you have made a sincere attempt to proactively solve problems.

Once your company representative is inside the customer's location, they should accomplish the following objectives:

1. Visit the point of use of the product. Talk with customer employees to solicit any information in relation to the product. Typically, when a customer contacts you about a problem with your product, it was an employee who uses the product in their job that originally complained. By talking directly with the employees, you may learn of a problem they are having that has not yet become a concern to their superiors and Supplier Quality yet. As well, by taking action on their concerns, they may be more likely to await your next visit rather than "cause an immediate commotion" over nonurgent problems that arise later. We have all received a phone call that resulted in same-day flights to the customer because one product out of 20,000 shipped was unacceptable. Establishing good relationships directly with customer employees who use your product will definitely pay off when the employee decides (because of having confidence in your company) that he'll just "hold on" to that one product until your next visit.

2. Visit all "holding" or quarantine areas where any of your company's accumulated rejects may be stored. Accumulations are sometimes held for months before your company is contacted. If product is found, you may be able to request permission to take it with you for investigation purposes. Also take advantage of this part of your visit to request samples of other supplier's products that mate to or are used with yours. Mating products are excellent sources of information and can often be obtained from the customer.

3. Visit the employees who are responsible for inspecting your company's product(s) upon receipt. Many times they can lend insight into potential problems. You may also find out if there are any characteristics or standards being inspected there that are not considered significant per the inspection criteria at your company. There is also the possibility that, though the criteria is the same, certain testing methods differ. If that is the case, copying the method used by the customer will assure compliance to that criteria.

4. Visit the warehouse area to see how your packages are handled and stored. You may notice that handling or storage methods

could affect the quality of the product packaged inside. Potential packaging changes could make the container more robust so that the customer's handling and storage methods would not cause damage.

The procedure that describes this overall process should also include the reporting method upon returning from the visit. You may decide to assign a customer concern serial number to any concerns found during the visit. This would insure that the concern will be treated and reacted to in the same manner as if the customer had contacted you with the problem.

9.3.3   Helpful Hints

Always follow up with a report to the customer with your findings during the visit and action plans for any concerns that were found. Your visit can become useless if you detected problems but did nothing to correct them.

When visiting with customer employees, take detailed notes that include their name, a unique description of what they looked like, and something personal that they talked about during the visit. You will significantly strengthen the relationships you are trying to build if you can approach an employee you met two months ago and can say, "Hi Greg, how are you doing today? By the way, how is your son Kyle doing after getting those tonsils out?" Taking notes to remind you of details like this can be invaluable. And the next time Greg gets frustrated over the quality of one of your products, he may be a little more understanding and patient.

Exhibit 29: Problem solving: Steps for "out-of-control" conditions on SPC charts

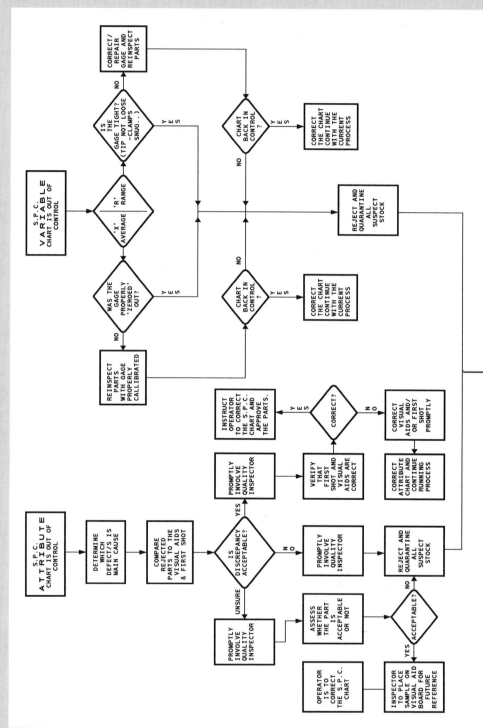

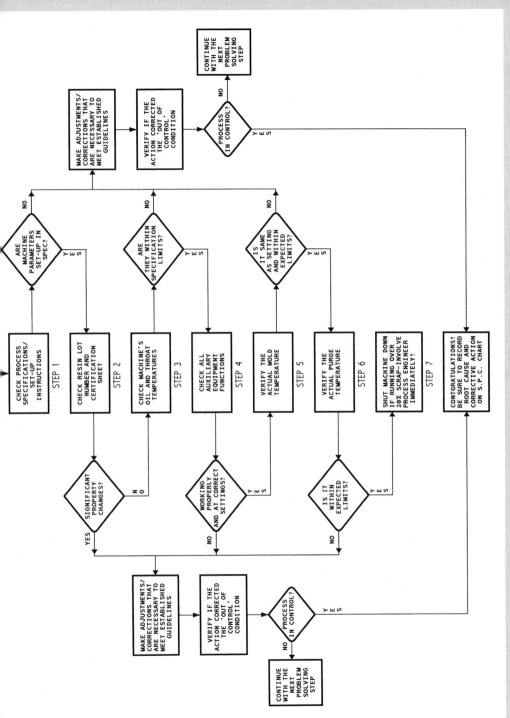

#	DATE	PART NUMBER/ DESCRIPTION	AMOUNT	CONCERN	TRACEABILITY: PackingSlip/Label	CONTACT/ PHONE #	CUSTOMER/ REF. #	INTERIM ACTION/S	TEAM CAPTAIN	PROJ./ACTUAL CLOSED DATE
				REJECT____ ACCUM.____ CONCERN____	PRODUCTION DATE- OPERATOR- SHIFT-					PROJECTED- ACTUAL-
				REJECT____ ACCUM.____ CONCERN____	PRODUCTION DATE- OPERATOR- SHIFT-					PROJECTED- ACTUAL-
				REJECT____ ACCUM.____ CONCERN____	PRODUCTION DATE- OPERATOR- SHIFT-					PROJECTED- ACTUAL-
				REJECT____ ACCUM.____ CONCERN____	PRODUCTION DATE- OPERATOR- SHIFT-					PROJECTED- ACTUAL-
				REJECT____ ACCUM.____ CONCERN____	PRODUCTION DATE- OPERATOR- SHIFT-					PROJECTED- ACTUAL-
				REJECT____ ACCUM.____ CONCERN____	PRODUCTION DATE- OPERATOR- SHIFT-					PROJECTED- ACTUAL-
				REJECT____ ACCUM.____ CONCERN____	PRODUCTION DATE- OPERATOR- SHIFT-					PROJECTED- ACTUAL-

Exhibit 30: Customer concerns tracking log

#: _____

DATE: _____

PART NUMBER: _____ RECEIVED BY: _____

PART NAME: _____ PLANT RESPONSIBLE: _____

CUSTOMER: _____

TYPE OF CONCERN: ☐ REJECTION ☐ ACCUMULATION ☐ COMPLAINT

CONTACTED BY: _____

PHONE NUMBER: _____ CUSTOMER REF.#: _____

PART NUMBER/S (IF MORE THAN 1)	AMOUNT RECEIVED	AMOUNT DEFECTIVE	PRODUCTION DATE	SHIFT	OPERATOR	WRITE **SCRAP** OR **RETURN** IF AUTHORIZED

* R E Q U E S T   T H E   A B O V E   I N F O R M A T I O N   I F   N O T   P R O V I D E D !

Concern Description: _____

_____
_____
_____

ROOT CAUSE/S: _____

_____

IMMEDIATE PLANT/ENGINEERING CONTACT (ROOT CAUSE/S INFO.): _____

Interim Action/Investigation

QUALITY MANAGER:

DO THE INSPECTION INSTRUCTIONS & OPERATOR IMSTRICTOPMS ADEQUATELY PROVIDE CRITERIA THAT, IF FOLLOWED, WOULD HAVE DETECTED THIS PARTICULAR CONCERN? ___YES ___NO * IF NO, EITHER UPDATE THE INSPECTION INSTRUCTIONS IMMEDIATELY AND DISPOSE OF OBSOLETE ONES OR ASSIST MNF. IN DOING THE SAME WITH THE OPERATOR INSTRUCTIONS.

ARE THERE ADEQUATE VISUAL AIDS DEPICTING THE DEFECT LOCATED AT THE POINT OF OPERATION? ___YES ___NO ___N/A *IF NO, IT IS YOUR RESPONSIBILITY TO WOTK WITH MANUFACTURING TO PROVIDE THE VISUAL AIDS NEEDED.

IF THIS CONCERN INVOLVES CRACKING, BREAKAGE, ETC. THAT MAY BE DUE TO THE RAW MATERIAL, TRACE THE RAW MATERIAL LOT NUMBER AND RECORD IT AS WELL AS THE TYPE OF MATERIAL UNDER "PROBLEM DESCRIPTION." ALSO, AS SOON AS SAMPLES ARE OBTAINED, OR REMAINING RAW MATERIAL FROM THE SUSPECT LOT, BEGIN TESTING IMMEDIATELY.

IS THERE ANY SUSPECT STOCK IN-HOUSE? ___YES ___NO IF YES, HAVE IT REJECTED AND QUARANTINED IMMEDIATELY!

IS THERE SUSPECT STOCK IN TRANSIT? ___YES ___NO IF YES, HAVE IT REVERSED OR CALL THE CUSTOMER IMMEDIATELY!

DID THE CUSTOMER REQUEST AN 8-D? ___YES ___NO IF YES, YOU ARE TO INITIATE IT IMMEDIATELY WITH THE PROBLEM DESCRIPTION, ROOT CAUSE/S, AND INTERIM ACTIONS. WORK WITH THE MEMBERS OF MANUFACTURING AND/OR ENGINEERING THAT ARE BEST ABLE TO PROVIDE THE EXPERTISE NEEDED TO TAKE ACTION/PREVENTATIVE MEASURES.

ANY OTHER COMMENTS OR OBSERVATIONS: _____

Q.A. MANAGER: _____ DATE: _____

CC: MOLDING MANAGER, MOLDING SUPERVISOR, SHIPPING, SALES, PLANT POST-BOARD, ENGINEERING, PRODUCTION CONTROL, ACCOUNTING

Exhibit 31: Customer concern notice (issue within one hour of receiving concern call)

```
ABC PLASTICS, INC.
1234 N. EASTERN HIGHWAY                                PART NUMBER:_____
KALAMAZOO, MICHIGAN  48756   PHONE: (810) 567-8923     DESCRIPTION:_____

CUSTOMER REFERENCE NUMBER:_____        CUSTOMER:_____

SUPPLIER CONCERN NUMBER:_____        DATE OPENED:_____
_____

TEAM MEMBERS:_____
_____


PROBLEM DESCRIPTION:_____
_____


ROOT CAUSE/S:_____
_____


INTERIM CORRECTIVE & CONTAINMENT ACTIONS (INCLUDE DATES):_____
_____


PERMANENT CORRECTIVE ACTION/S (INCLUDE DATES):_____
_____
_____


VERIFICATION OF ACTIONS' EFFECTIVENESS (USE MEASURABLE):_____
_____


PREVENTION OF RECURRENCE:_____
_____

CONGRATULATIONS TO THE TEAM UPON CLOSING THIS REPORT!

REPORT PREPARED BY:_____      DATE:_____
```

Exhibit 32: Preventative action report (form)

10 Nonconforming Material Control

10.1 In-House Rejections

The acceptance criteria for products should be "zero discrepancies." This means that one or more discrepancies would result in the lot, batch, etc. being rejected. Rejection procedures should be documented, and outline all steps from the point that a product is deemed unacceptable to the point that the product is made acceptable and approved, scrapped, or ground for reuse of the material. In this portion of this chapter, the focus will be specifically on rejection, reporting, action, and tracking methods. Returned goods, quarantine, and sort/rework operations will be discussed in more detail later in this chapter.

10.1.1 Different Stages of In-House Rejections

At each phase of inspection is the opportunity to either accept or reject the product as meeting or having not met the inspection criteria. Those phases include:

Inspection phase	Can result in the rejection of:
Start-up inspection	The process
In-process inspection	All production to the last point of in-process approval
Final inspection	The entire carton/batch/lot that the sample inspected represents
Dock-audit inspection	The entire lot that the sample inspected represents and any other suspect stock
Annual validations	All in-house stock is "suspect"

Procedures should indicate exactly what and how much should be rejected if the inspection of a sample results in detecting an unacceptable product.

10.1.2   The Rejection Process

It is crucial that the rejection process provide a means of rejecting stock, reporting, taking action, and tracking. To accomplish all of these steps, requires the preparation for them at the first step—rejection.

Initially, in the event that stock is being rejected, the suspect stock is to be identified as "rejected" and immediately segregated from the normal flow of production and/or storage. Typically, rejected stock is identified with a label, tag, stamp, etc., that is a specific color (red, orange, etc.) so as to be easily recognized as "defective." The identification label should contain the product number, description, quantity, date and reason for rejection, inspector identification, etc. To facilitate the remaining phases of rejecting stock, the following components should be considered:

1. Assign a serial number to the rejection. The serial number can be applied to all related documents, including each individual rejection identification label. The number will tie all pertaining information and stock together.
2. A report should be initiated that describes rejection details, sort/rework/scrap details, the action plan, etc.
3. Start a tracking log to house the assigned serial number, rejection identification label information, and the closed date (date that the report is finalized).

10.1.3   Reporting and Action Plan Documentation

A report of the rejection should be initiated and contain at least the following information:

- Product number and description,
- inspector identification and recommendation (sort, scrap, etc.),
- date and time of the rejection,
- reason for the rejection,

- total quantity rejected and the percent defective, and
- traceability information (machine number, production dates, shift, resin lot number used, etc.).

The initiated report should immediately be copied and issued to all departments. The report can be used later by manufacturing to record the following interim and permanent action information:

- Disposition (scrap, rework, sort, etc.),
- sort/rework details (employee assigned, method, quantity/percent reworked/sorted out, amount of time to perform rework/sort, and the inspector identification who approved the stock afterward), and
- action plan details (root cause, actions taken, prevention, verification, approval).

Once the manufacturing department completes the report, it should be forwarded to Quality for tracking purposes and filed in the job file. Refer to Exhibit 33 for an example of a rejection report.

10.1.4   Tracking and Improvement Efforts

Tracking is essential as a resource for reference. It provides the data necessary for analyzing statistics in order to further improve overall processes. The tracking log, at a minimum, should contain the basic information in regard to each rejection and the dates that reports were opened and closed.

Many topics relating to the rejection process can be charted to assess improvements in manufacturing processes and the overall rejection process. The following list is of some example topics that can be measured and tracked at given frequencies (typically on a monthly basis) and reviewed by a team for improvement efforts:

- Time turn-around of rejection reports: On a monthly basis, the average number of days it takes to close a rejection report can be calculated and recorded on a bar or line chart.
- Average number of closed reports: On a monthly basis, the percent of closed reports can be calculated (based on amount issued and amount closed) and charted.

- Major defect focus: A tally sheet can be kept that lists each type of defect and provides space to record the quantity of parts rejected under the pertaining defect. Based on actual numbers or percent of the overall accumulation, a Pareto chart can be made.
- Major product line focus: In the same manner as above, a Pareto can be made to provide insight into priorities.
- Reject rate:The average percent of rejects can be calculated from production quantities and charted to assess performance.
- Rework/sort dollars and time: Time spent on (and costs associated with) rework and sort operations can be compiled and charted for accountability.

Many topics can be combined on one chart. For example, rejection report time-turn-around and percent of closed reports per month can appear on a combination graph as shown in Exhibit 34.

Goals for each measurable can be determined based on history and expectations to develop an annual improvement plan. Actions can then be taken on a larger scale to reach the goals.

10.2   Returned Goods

Product that is returned from the customer should be subjected to your company's rejection procedure in regard to identifying it as "rejected" and segregating it for review. Notification that product is being returned is usually received before the goods are returned. If so, the stock will have already had a customer concern serial number assigned to it (if not, one should be assigned). Once the returned goods are received and identified as "rejected" (record the customer concern serial number on the label), segregate and contain the material. The process for addressing customer concerns should apply to this activity (refer to "Customer Concerns" in the chapter on "Problem Solving").

10.2.1   Customer Concerns Process for Returned Goods

Returned goods should be evaluated to aid in root cause analysis. If the root cause has already been determined, the returned product should still be reviewed to verify any discrepancy reports. The percent defective should be determined and the possibility of rework investigated. The information gathered during the review of the stock should be recorded on the problem solving report under the pertaining areas, ("problem description," "root cause," and "immediate actions"). The rest of the process for addressing customer concerns should continue as outlined in your company's procedure.

10.2.2   Disposition of Returned Goods

After returned goods have been evaluated, a disposition should be made to sort, rework, scrap, or grind the products. The disposition should be recorded in the related customer concern record and on any other document(s) where procedures apply. The disposition should then be executed.

10.2.3   Sort/Rework Operations on Returned Goods

If sort or rework operations occur on returned goods, the sort/rework procedures should indicate how information with respect to those operations are to be recorded for tracking purposes. New shipping labels should be applied to the container(s) and they should be reinspected and approved according to the inspection criteria. Packing slips or labels should contain the information necessary to relate back to the original production dates of the product.

10.3   Quarantine Procedures

All discrepant stock that is rejected in-house or returned from the customer should be immediately segregated directly after being identified as "rejected." The only way to assure effective segregation is to have a designated quarantine area situated away from the normal flow of production and storage. This process should be documented in rejection procedures (a flow chart is an excellent visual aid that should accompany a procedure that affects the flow of production/stock).

10.3.1   Quarantine Area Location and Design

The quarantine area should be located in a position in the plant that is away from the opportunity for mistakes. It should not be situated, for example, anywhere near production areas, finished goods storage, or shipping. It should not be located anywhere within the traffic flow from production to finished goods storage/shipping. Some good places to station the area would include a corner or back wall in the "in-process" storage area or "receiving" area.

To further reduce the possibility of rejected stock becoming mixed with acceptable stock, the following measures should be seriously considered:

1. Label the quarantine area with signs, outline the area on the floor with painted lines, etc., to assure it is easily recognized as the quarantine area.
2. Close off the area by surrounding it with a fence, partition, or walls.
3. Provide locks for the area and issue keys only to those personnel who need access to the area.

10.3.2   Quarantine Contents and Dispositions Tracking

Records should be maintained on all items that enter the quarantine area. The record should display the:

- Date that the goods were entered,
- name of the employee who entered them,
- reason they are quarantined,
- disposition,
- date of release, and
- name of the employee who removed them.

The record would be best kept in the form of a log sheet that remains at the entrance to the quarantine area. Refer to Exhibit 35 for an example log sheet.

An individual should have the assigned responsibilities for assuring that the contents of the quarantine area have been properly accounted for. As well, that employee should police the timely disposition of stored goods, and keep the area organized.

10.4   Sort and Rework Operations

"Rework" is a term used in many plastics companies and in a lot of cases, refers to secondary operations. In fact, the actual "reworking" of product is typically minimal in many plastics operations. It is usually only performed as an interim action while permanent design changes or tool corrections are in progress. For this reason, it is important to define rework and other types of status that result in a product temporarily being unapproved for shipment.

10.4.1   Different Definitions of Stock Status

Rework: Product, that for any reason, requires a secondary operation that is not shown on the process flow chart and is not considered "normal." For example, the drilling of an extra hole in the existing product for whatever reason, would be considered rework. Rework should always be approved by the customer because it can result in the product varying from the current blueprint.

Unfinished: Product that (for any reason) was not subjected to the normal process flow, thus resulting in the product being unfinished. For example, gates have not yet been removed from the production.

Sort: Production that is suspected to contain a certain percent of discrepant product. For example, a carton was rejected because a short shot was found during final inspection. The entire container would have to be sorted to segregate the unacceptable from the acceptable product.

Repack: Production that needs to be repackaged for any reason. For example, a damaged carton or the use of a temporary carton would need to be repackaged into acceptable packaging.

10.4.2   Rework/Sort Area and Instructions

Rework, sort, etc., operations should only be performed in a designated area. That area should be located away from the production process and near the quarantine area. The area should be well lighted and provide the work space necessary to perform an organized operation.

Written and visual instructions should be posted at the rework, sort, etc., point of operation. The operation should be treated as a regular operation process. Quality and Manufacturing should insure that the employee who is responsible for the operation is performing it according to instructions and expectations. The operation should then be revisited at the same frequency that normal in-process inspections occur to verify that it is producing acceptable product.

10.4.3   Rework/Sort Records and Inspection

Records should be maintained on rework/sort results, traceability information, and inspection approvals. Refer to the "rejection report" referenced in the "In-House Rejections" section of this chapter for an example.

During the operation, the operator should keep track of and record the number of parts reworked/sorted, the quantity/percent defective and reason(s), and the amount of time spent on the operation.

Either through records, a bar-coding system, labels, packing slips, etc., the reworked/sorted product should be traceable back to the original production date. It is important that the product be recognized as having been reworked/sorted on the date that the operation occurred.

Inspection should occur according to regular inspection procedures. A "first piece approval" should be displayed at the operation and routine inspections should occur at normal frequencies thereafter. The finished product should then be subjected to final and dock audit inspections as if it were "virgin" material. In other words, the finished production should be subjected to all inspection criteria, regardless of the fact that it was sorted for "splay" only. The reason a policy such as this should be adopted is because an inspector will inherently discontinue the inspection if discrepant goods are detected. In other words, rejected stock could have not been subjected to a full inspection of all criteria at the full sample size. Even if it was fully inspected, the goods could become scratched, dirty, etc., during the sort operation. At the very least, a cursory check of all other criteria unrelated to the rejection should be checked on a small sample of the stock. If the finished product is once again found to be unacceptable, it should again be rejected according to regular rejection procedures.

SERIAL # 2546

PRODUCT NUMBER OR MATERIAL: _____ SUPPLIER (IF APPLICABLE): _____

PRODUCT DESCRIPTION: _____ DEPARTMENT RESPONSIBLE: _____

TOTAL QUANTITY DEFECTIVE: _____ NUMBER OF CARTONS: _____ TOTAL QUANTITY PER CARTON: _____

LOT NUMBER: _____ DATE ISSUED: _____ ISSUED BY: _____

DEFECT OR DISCREPANCY: _____

QUALITY ASSURANCE RECOMMENDATIONS: _____

UPON COMPLETION OF ABOVE PORTION, DISTRIBUTE COLORED CARBON COPIES AS INDICATED

THE FOLLOWING SECTION IS TO BE COMPLETED BY THE DEPARTMENT/SUPPLIER RESPONSIBLE:

DESCRIPTION OF REWORK/SORT TECHNIQUE: _____
_____

DISPOSITION OF PRODUCT/MATERIAL: _____
_____

CAUSE OF DEFECT/DISCREPANCY: _____
_____

PERMANENT CORRECTIVE ACTION: _____
_____

SIGNATURE & TITLE: _____ DATE: _____

THE FOLLOWING SECTION IS TO BE COMPLETED BY THE REWORKER/SORTER:

QUANTITY SCRAPPED	REWORK/ SORT DATE	REWORKER/ SORTER	REINSPECTED BY:	REINSPECTION DATE	QUANTITY RETURNED TO STOCK	TOTAL REWORK/ SORT HOURS

FINAL DISPOSITION   SERIAL # 2546

PRODUCT/MATERIAL NUMBER: _____ QUANTITY RETURNED TO STOCK: _____

DETACH BOTTOM PORTION UPON COMPLETION & FORWARD TO INVENTORY

Exhibit 33: Rejected parts status report (form)

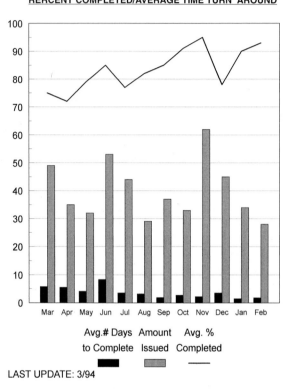

RERCENT COMPLETED/AVERAGE TIME TURN AROUND

Avg.# Days   Amount   Avg. %
to Complete   Issued   Completed

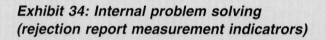

LAST UPDATE: 3/94

*Exhibit 34: Internal problem solving
(rejection report measurement indicatrors)*

REJECTION/ HOLD REPORT SERIAL NUMBER	MATERIAL I.D./ PART NUMBER	DESCRIPTION	QUANTITY ENTERED	DATE ENTERED IN QUARANTINE	REASON FOR QUARANTINE	FINAL DISPOSITION	DATE RELEASED	INSPECTOR RELEASING

Exhibit 35: Quarantined material – status log

11   Lot Control and Traceability

11.1   Lot Control

Incoming goods, raw materials, produced goods, in-process goods, and approved/finished goods should each be identified as to their status in a way that is easily recognizable. For example, label colors may differ according to status: pink labels for incoming goods, blue labels for in-process, white labels for finished goods, etc.

Each type of material should be stored in a designated area that is separate from each other. The storage areas should also be marked so that they are easily identified as the area for storage that they represent. No stock should be moved into its storage area without being marked as "approved" by Quality.

All material that is rejected or "on hold" should be identified as such and stored in the quarantine area.

11.1.1   Documenting the Lot Control System

Every plant will have its own system of lot control. The basic task is to assure that every classification of product/goods has its own identity and is stored in its own area. Once the lot control system is developed, it should be documented. I suggest that a flow chart is best used to record and depict this process. Refer to Exhibit 36 for an example of a Lot Control Flow chart that documents the lot control system at a small "shoot and ship" molding plant.

11.1.2   Lot Control Further Defined

Lot control can be further defined in terms of organizing the system and considering such quality aspects as shelf life. "First in-first out" (FIFO) storage methods assure that product is shipped in the order that it was produced. In addition to the obvious reasons for "first in-first out" lot organization is the enhancement of traceability efforts.

"Min-max" storage processes facilitate low inventory efforts by assuring that only enough material for the next run is in storage at all times. This system would require daily inventory verification by visually checking for empty specified storage locations of in-process materials.

Though product shelf life is not a major concern for the finished goods of plastic manufacturers, consideration should definitely be given to certain materials and assemblies. The length of storage for glues, self-adhering pads, paint, etc., should be monitored as well as those finished products using those types of materials.

11.2   Traceability

Defined here, traceability is the ability to trace a finished product back to the exact raw materials, machines, personnel, and processes that made it. The purpose for traceability is usually to retrieve all related records/samples for the timely problem solving necessary in the event of a quality concern. It is when the finished product consists of many components and materials that the process of assuring traceability will become very intricate and more difficult.

11.2.1   Traceability Methods

For a company that produces products made from just one material, the task of recording the resin lot number on the process record or inspection record will suffice for tracing purposes. As long as the production date on the container's label can be identified in conjunction with the product it contains, the resin's lot number should be easily attained.

If the product should become separated from its container, traceability will be lost unless each individual product is identified with traceability information. One method for individual product identification is the "date wheel." Date wheels come in many forms and are located in molding tools (or dies). Typically, they can be updated easily with a "punch" or by turning an insert. At a minimum, date wheels show the year and month that the product was produced. Later models go as far as to indicate the exact date and period of time in which the part was produced.

When three different molded components and three different purchased components (i.e., spring, pin, and adhesive pad) make up one assembled product, the task of traceability becomes very elaborate. Every component will have to be marked with traceability information, or a record must exist for the finished product that details lot information for each component. Because some products are simply too small to be identified individually, it makes sense that a record should exist that houses the origin of all components.

Today, the best technology available to augment a complex traceability system is bar-coding. It is the design purpose of bar-coding to track and record information efficiently and effectively. A process using a bar-coding system would involve "reading" the bar-code on the carton of each component with a "wand" and expanding the bar-code file to envelope those serial numbers that make up the finished product.

A manual method for tracing origin of all components of an assembly is the "traveller" system. This involves recording the lot number of all used components on a tag that travels with the product through each stage of its assembly. The tag is then labeled with the final production date and filed for reference while a copy of it is placed in the finished carton.

11.2.2   Documentation of the Process

Once the process is determined, the details of the system should be outlined in a procedure. Again, a flow chart that depicts the system is an excellent visual aid to assure execution of the process.

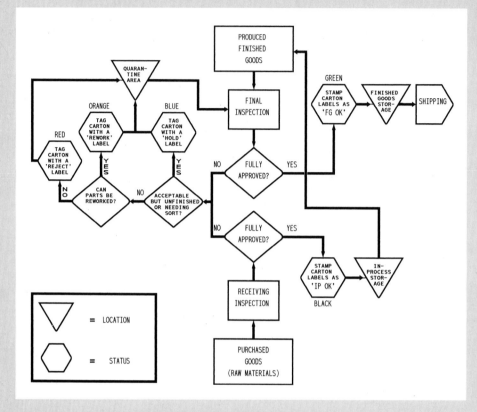

Exhibit 36: Lot control system (plastic molding facility)

12   Document Control

12.1   Record Retention

A record retention policy and related procedures should assure the following benefits:

1. All records needed for future reference will be available when needed.
2. All stored records will be organized, identified, and kept safe in a designated area to ensure ease in their recovery.
3. Any customer and governmental expectations for record retention can be met through the record retention policy.
4. Documents not needed (or no longer needed) for future reference can be disposed of in a systematic manner.

12.1.1   The Policy

The record retention policy, like all policies and procedures, should be a living document that is consistently updated and improved. To begin, a multi-disciplined team should consider any customer and governmental requirements in addition to current needs and develop retention standards.

The following is an example of the requirements that may be listed in a record retention policy:

- All records relating to saleable products that are considered "safety-related" will be retained for a period of five (5) years.
- All records of written authority to vary (even temporarily) from product engineering specifications will be retained for a period of five (5) years.
- All sample submission records and approvals are to be retained for the life of the product plus two (2) years after obsolescence.

- Any operator/inspector instructions that are obsolete due to revisions must be kept for six (6) months.
- All other records to be retained for a period of two (2) years.

12.1.2   The Procedure

Record retention procedures will vary from company to company. However, provisions should be included in each procedure to address at least the following areas:

- All files and records should be compiled for storage at certain intervals. Some may be stored monthly while others need only be collected annually.
- Responsibility should be defined for all steps of the process.
- Directions for storage should be outlined that describe packaging and storage location.
- Details on package identification should require at least the following information:
  Description,
  department,
  date stored,
  storer's name, and
  date that the contents can be disposed of.
- Requirements for the performance of audits that are to occur at established frequencies to dispose of all records that have exceeded storage time requirements.

Make sure that stored records will be properly identified with all necessary information. Poorly identified containers will result in lost records as well as a "pile up" of records that should have long ago been purged from storage. A pre-made self-adhesive label that describes the information needed and provides space to record it will lessen the chances for poor identification of stored records. Refer to Exhibit 37 for an example label.

12.2   Procedure and Instructions Revisions

All written procedures and instructions should be identified by the date that they were released and the date that they were last revised. A revision record should be kept that tracks all revisions made and describes the extent/effect of any changes. All revisions should be approved by a designated individual of management. Obsolete procedures and instructions should be removed from the system. A process to accomplish these requirements should be documented and include any other company guidelines with respect to revising/updating procedures, instructions, policies, etc.

12.2.1   Reasons for Revisions/Updates

Customer or in-house specification changes, (dimensional, material, testing, identification, packaging, etc.) and improvements of methods resulting from problem solving efforts, action plans, etc. are typical causes for changing instructions. Procedures and policies are usually changed for similar reasons. As well, changes will occur as a result of system enhancements, new technology, new customer expectations, and to fine-tune written processes to more factually correlate to actual processes.

12.2.2   Revisions Tracking and Control

Special emphasis should be given to ensuring that revisions do not result in more than one version of a procedure or instructions in circulation at the same time. The system should address exactly how obsolete procedures are removed from the system and disposed of or stored.

In addition to a "revisions summary" that tracks and details all revisions, there should be a "master" file or book that houses all procedures, instructions, policies, etc. The master file should have designated location(s) and an individual assigned with the responsibility for maintaining the file(s).

All procedures should be reviewed at least annually to assess the need for revisions, improvements, and updates.

12.3   Work Station Documents

One of the best ways to facilitate the process of procedures and instructions being read is to post them at the point of use. Because the majority of the work force in any manufacturing company is located at various stations on the manufacturing floor, focusing the "posting" efforts in this area may yield the greatest results.

12.3.1   What Should be Posted

All procedures, visual aids, and instructions relating in any way to the responsibilities of the manufacturing personnel should be easy to access by those personnel. However, specific procedures and visual aids should be posted/displayed at each work station as an immediate resource for the employee stationed at an operation. At a minimum, I suggest that the following items be displayed:

- First approved sample (from the beginning of the run),
- visual aid board (depicting acceptable/unacceptable product),
- operator/process instructions (pictures, diagrams, and flow charts increase information retention),
- process set-up instructions and work station lay out,
- inspection instructions (visual aids can be used here also),
- any job specific gaging and SPC instructions, and
- all customer concern/rejection reports.

Charts and graphs that show progress of scrap rates and capability studies should be posted if possible. Not only will these types of "history sheets" provide good information while problem solving, but they make it clear to the operator that data is used and evaluated, and actions are taken.

12.3.2   Standardization of Work Station Documents

Standardizing the placement of documents and visual aids at the work station will provide a uniform presentation of the information. It will also aid in reminding those responsible for posting information, of exactly what is to be displayed at the station so that there is consistency.

Standardizing a work station will necessitate reserved space for the documents/visual aids. This could be accomplished by attaching a peg board to the work table or to a free standing, moveable podium. Clips could then be attached in specific areas on the board for procedures and visual aid boards to be added. Each area on the board could be labeled to identify what that space is reserved for.

Once standards are developed for the display of visual aids and procedures, the process and an outline of the standards should be documented. Refer to Exhibits 38 and 39 for examples of work station visual aid standards.

12.4   Files

Unorganized files can prove useless. All documents should be filed according to a system that assures organization and easy retrieval. Files are typically set up by program or job in numerical or alphabetical order with each job file housing all related documentation. Some files are set up by type of document in order of the job numbers, (i.e., one file drawer for FMEAs, one for control plans, one for inspection instructions, one for flow charts, one for layouts/blueprints, etc.). I do not recommend the latter because ten different files may need to be accessed to retrieve all documentation for one job number.

12.4.1   Color Coding

Color coding files allows for all documentation relating to one job to be easily recognized in categories. For example, store all control plans and flow charts in green folders (or label the folder with a green pen), all blueprints and layouts in blue folders, all inspection instructions in red folders, all obsolete documents in black folders, all SPC documentation in yellow, etc. This method is the beginning of creating a file system.

12.4.2   Organizing a Filing System

As suggested, all files should be placed in numerical or alphabetical order according to the job or program identification. There may be separate file cabinets for each major customer or to separate different processes: molded product numbers, assembled product numbers, painted part numbers, etc. If there are two or more identification numbers or names for the jobs/program (i.e., customer part number and in-house I.D. number), there should exist a correlation chart that is kept near the files for reference.

A "master" file list should exist that outlines all major job files and the latest engineering change level. This master list could be tied in with a job number correlation list if one exists. The master file list should be updated when a new file is added or the engineering change level is updated.

12.4.3   File Audits

A procedure could be written that assigns responsibility for file maintenance and requires audits to be performed at established frequencies. During the audit process, all contents containing the blueprint's latest engineering level should be verified to the master file list. Also, misfiled, obsolete documents should be removed and any missing documents replaced. A copy of the master list could serve as a checklist during the file audit.

12.4.4   Helpful Hints

Do not assume that documented procedures are not necessary for a filing system. It is very important that all persons responsible for maintaining the files are designated in a procedure. Procedures should also provide guidelines for anyone who might need access to the files so that they understand how to retrieve a file and *return it.*

Post a sign on or near the file cabinets that informs employees that all contents of the file are to *remain* in the office in which the files are located. Provide a "check out" log and require that anyone wishing to "borrow" a file signs for it before removing it from the office.

RECORDS STORAGE LABEL

CONTAINER CONTENTS:

DATE STORED:

DATE TO BE DISPOSED:

STORER'S NAME:

*ADHERE THIS LABEL TO THE
CONTAINER THAT IS BEING
STORED.*

Exhibit 37: Simple storage label

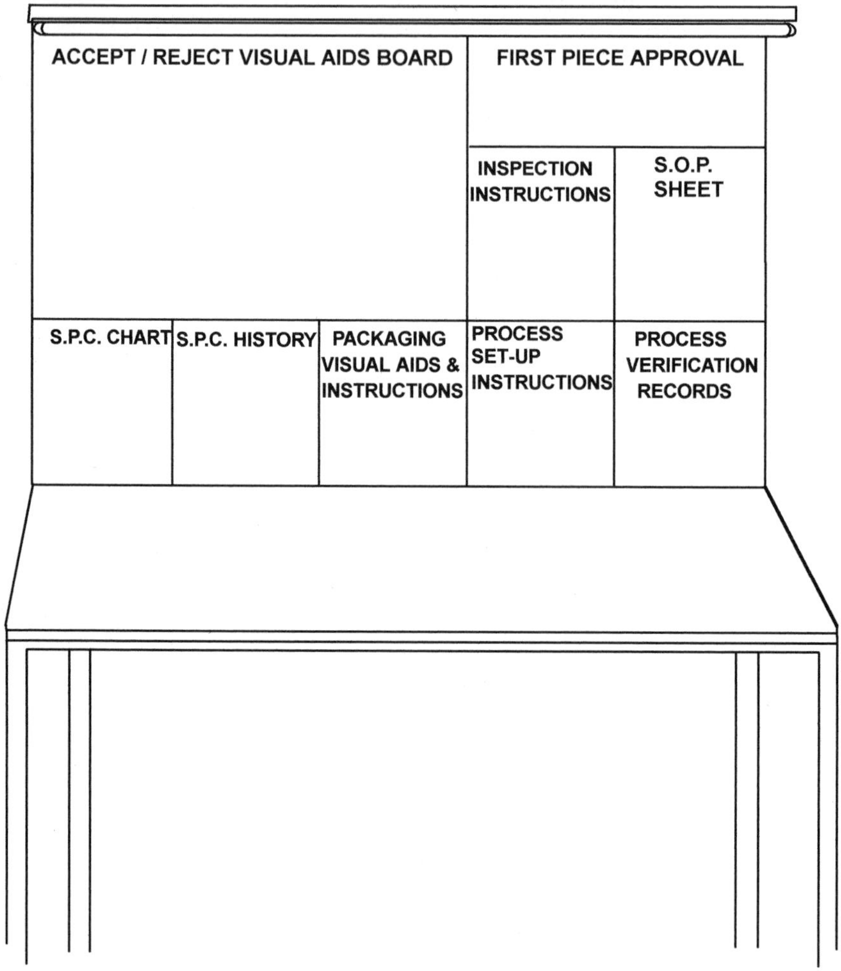

Exhibit 38: Work station visual display standard (work table)

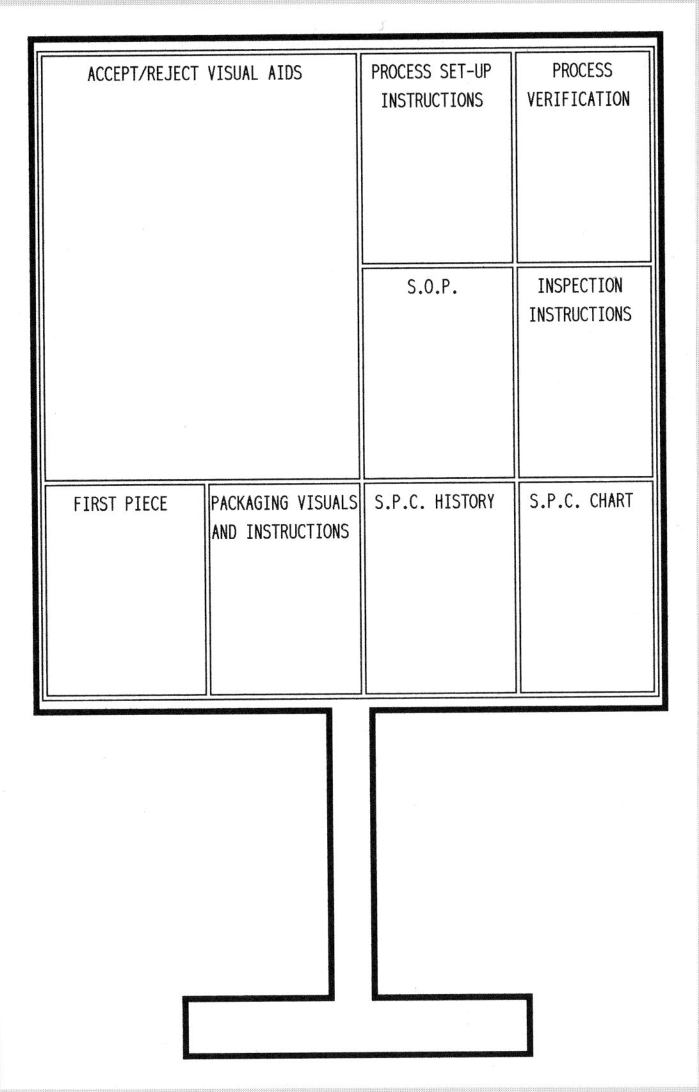

Exhibit 39: Work station visual display standard (podium)

13 Internal Systems Control

13.1 Self-Assessments

In order to understand how your company is performing to internal and external requirements and procedures, formal assessments should be conducted at least quarterly. These self-assessments will provide the company with:

1. Detailed information to measure improvements in each area,
2. a real understanding of where requirements are being met and which procedures are properly followed, and
3. the direction needed to insure improvement in all areas where procedures are not properly followed, requirements not met, or if either did not result in the benefits expected.

13.1.1 Assessment Contents and Procedures

The assessment should provide for investigation into all key requirements and internal/external expectations of your company in all areas: quality, manufacturing, technology, delivery, finance, and leadership. Please refer to section 4.2, "Supplier Assessments" in Chapter 4, "Supplier Development And Control," for more on assessment contents and a numerical rating system. The rating system to be used should include explicit definitions for each type of rating.

If the assessment form just briefly describes requirements, an "assessor's guide" should be developed that details exactly what is expected for each requirement to be met. To conduct assessments in a consistent manner, the expectations must be objective and clearly outlined.

In addition to an "assessor's guide" to interpret the components of the requirements, a procedure should be documented that describes the assessment process. It should contain the purpose, frequencies, responsibilities, process, and follow-up.

13.1.2   Responsibilities and Scheduling

Establish an assigned responsibility to oversee the entire assessment process. Their duty would be to issue reminders, maintain all assessment results and action plans, and keep any other related records and tracking information. The person assigned this responsibility should have the authority to assure the process is followed. A Corporate Manager, the Controller, General Manager, or Human Resource Director are some of the individuals who may be given custody of the assessment program.

The most effective assessment will occur when conducted by someone who is "removed" from the department or on a corporate level within that department. I suggest that the same person not be responsible for conducting an assessment on the same area two times in succession. Those too closely related to the area are more likely to make assumptions and overlook important aspects in regard to requirements being fully met. Having different "eyes" conduct the assessment will provide more insight and ideas to better improve certain areas. Another way to vary assessors, is to have a team of individuals responsible for assessing each area.

Responsibility should also be assigned to an individual who is to serve as the main contact for each individual area. Typically, that responsibility would be that of the area manager. This person would also be responsible for assuring that action/improvement plans are developed, issued, and executed.

As mentioned above, assessments should be conducted on each area at least quarterly. A schedule should be prepared annually that lists the following:

1. Assessment dates,
2. area to be assessed,
3. assessor, and
4. contact person.

Refer to Exhibit 40 for an example of a one-year assessment plan schedule.

13.1.3   Action Plans

Action plans should be developed to correct or improve every area requirement that did not receive a perfect score during the assessment. Action plans will be most effective if the assessor documents all details of their findings during the assessment and includes constructive improvement suggestions. After the area team has documented the action plan, it should be published to the assessment program coordinator and the main contact. Refer to "Supplier Assessments" for details on constructing an action plan.

When the following quarter's assessment is being scheduled, the assessor for each area should obtain a copy of and follow up on the last quarter's action plan.

13.1.4   Reporting and Improvement Tracking

The results of all assessments and actions taken should be reviewed in a regularly scheduled staff meeting. It is also important to report how the results in each area compare with previous results.

If the assessment provides numerical ratings, line graphs can be maintained to depict the ratings history. Improvements, upward/downward trends, etc. could then be monitored visually. Refer to Exhibit 41 for an example of charted tracking of assessment scores.

13.2   Annual Product Validation

To maintain product quality according to blueprint specifications and standards, each product that your company produces should be validated annually. Usually, once a product is approved for production, only a few measurable characteristics are inspected on a routine basis. Product validation consists of inspecting the product to every blueprint and/or standard specification just as it was during the sample submission process (please refer to Chapter 3, "Sampling and Submissions," of this book, specifically Section 3.3, "Sample Submission To The Customer or Production Approval Source").

A list of active product numbers should be maintained along with the last annual validation date to serve as a checklist. A procedure should describe the process, responsibilities, and actions to be taken in the event of a failure/out of specification measurable. If a failure or unacceptable condition is found during validation, normal rejection procedures should be followed.

It should be the requirement of your company's suppliers to conduct the same verifications on products/material shipped to you.

13.3   Floor Audits

The conducting of daily floor audits on the manufacturing floor in a structured manner provides a certain "ritual" of follow-up. To begin with, the daily floor audit should be formal (written with premade forms for tracking and recording results; refer to Exhibit 42 for a daily floor audit form example), and contain important items that are currently part of manufacturing procedures. The items may be of extreme importance, safety-related, tool checklist items, etc. The audit might contain some of the following items to be checked:

1. Are all necessary documents posted at the work station? (i.e., inspection/operator instructions, SPC/gaging instructions, process set-up sheet, FMEA, SPC history sheet/graph).
2. Are gloves and safety glasses being worn by the Operator?
3. Is the work area clean and safe?
4. Has the Operator been provided all tools necessary to perform his job correctly and efficiently?
5. Is the approved "first sample" posted at the process?
6. Is the process being performed according to the flow chart?
7. Are there adequate visual aids for acceptance/rejection criteria displayed at the work station?

A floor person, supervisor, etc. should be assigned responsibility for conducting the audit daily on each shift. Any discrepancies should be corrected immediately or a written action plan should be developed to correct them as soon as possible.

Procedures should already exist that provide the processes for each item to be in place prior to the audit. In other words, the audit should not be the only means of assuring that the items are correct, but rather that the procedures are being followed. Therefore, actions taken as a result of a discrepancy should not focus only on correcting the item but investigating why the procedure was not followed and how to prevent reoccurrence.

13.4   Housekeeping

It is important to me that this topic receives its own section because it is a vital aspect of quality, organization, work environment, safety, employee morale, and overall company image. Good housekeeping should be a mission of every company. It should be mandated, advertised, encouraged, and consistently supported by all levels of management and supervision (not just when there is a customer tour). It should be expected of all management and supervisory personnel to set examples and not let their ego or "power" get in the way of stooping down to pick up a piece of trash if it's in their path.

No one can (or will) disagree with the above statement because it is true and rational. This is why it is difficult to understand why a large portion of plastic manufacturing shops appear to be in complete disarray. Please refer to Chapter 15 in this book on "Systems Implementation," specifically Section 15.1, "Top Management's Role," for an understanding of how poor housekeeping could exist.

Month	Area	Assessor	Contact
MARCH:	QUALITY	CONTROLLER	QUALITY ASSURANCE MANAGER
	COST	PURCHASING MANAGER	CONTROLLER
	DELIVERY	ENGINEERING MANAGER	PRODUCTION CONTROL MANAGER
	TECHNOLOGY	MANUFACTURING MANAGER	ENGINEERING MANAGER
	LEADERSHIP	QUALITY ASSURANCE MANAGER	PERSONNEL MANAGER
JUNE:	QUALITY	DIRECTOR OF OPERATIONS	QUALITY ASSURANCE MANAGER
	COST	PERSONNEL MANAGER	CONTROLLER
	DELIVERY	QUALITY ASSURANCE MANAGER	PRODUCTION CONTROL MANAGER
	TECHNOLOGY	PURCHASING MANAGER	ENGINEERING MANAGER
	LEADERSHIP	PRODUCTION CONTROL MANAGER	PERSONNEL MANAGER
SEPTEMBER:	QUALITY	PURCHASING MANAGER	QUALITY ASSURANCE MANAGER
	COST	ENGINEERING MANAGER	CONTROLLER
	DELIVERY	MANUFACTURING MANAGER	PRODUCTION CONTROL MANAGER
	TECHNOLOGY	PERSONNEL MANAGER	ENGINEERING MANAGER
	LEADERSHIP	DIRECTOR OF OPERATIONS	PERSONNEL MANAGER
DECEMBER:	QUALITY	ENGINEERING MANAGER	QUALITY ASSURANCE MANAGER
	COST	MANUFACTURING MANAGER	CONTROLLER
	DELIVERY	CONTROLLER	PRODUCTION CONTROL MANAGER
	TECHNOLOGY	QUALITY ASSURANCE MANAGER	ENGINEERING MANAGER
	LEADERSHIP	PURCHASING MANAGER	PERSONNEL MANAGER

Exhibit 40: Self-assessment schedule (one-year plan)

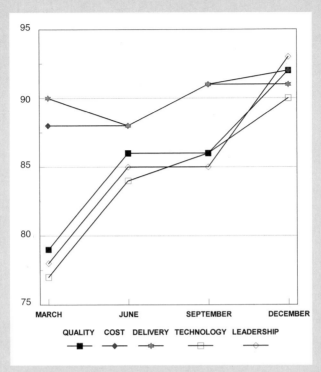

Exhibit 41: Self-assessment scores
(charted tracking)

DATE:_____ **CHECKLIST**

SHIFT:_____

CHARACTERISTICS TO BE ASSESSED	WORK STATION NUMBER										COMMENTS
	1	2	3	4	5	6	7	8	9	10	
JOB NUMBER:											
FIRST PEICE POSTED											
WORK AREA IS CLEAN & SAFE											
OPERATOR INSTRUCTION POSTED & REVIEWED											
INSPECTION INST.'S POSTED & REVIEWED											
PROCESS VERIFICATION POSTED & UPDATED											
S.P.C. POSTED & IN CONTROL											
DATE CODE ON PRODUCT IS CORRECT											
VISUAL AIDS ARE POSTED & REVIEWED											
OPERATOR HAS ALL NECESSARY TOOLS											
PROCESS PERFORMED ACCORDING TO POSTED PROCESS FLOW CHART											
CORRECT GAGE IS AT MACHINE & WORKING											
OTHER (EXPLAIN)											

* ALL DISCREPANCIES MUST BE CORRECTED IMMEDIATELY!

SUPERVISOR SIGNATURE: _____

Exhibit 42: Daily floor audit form

14   ISO-9000

It is important for me to include a chapter on ISO-9000 in this book because: 1) It can be a very valuable quality system tool, and 2) I believe anyone responsible for a quality system anywhere in the world should at the very least be familiar with what ISO-9000 is about. The ISO-9000 series includes standards that may be used for quality management guidance and identifies certain quality system elements that are necessary to achieve quality assurance. The series does not constitute a total quality management system; rather, it can be used as a model to construct the quality system foundation for total quality management, various quality awards, or specific customer quality requirements.

14.1   The Origination and Purpose of ISO-9000

The International Organization for Standardization (ISO) was founded in 1946 to develop a universal collection of manufacturing, trades, and communications business standards. ISO is based in Geneva, Switzerland, and is composed of 91 member countries. In 1987, a series of quality standards was published by ISO called the "ISO-9000" series. Unlike the United States, whose quality standards are mainly customer originated and enforced, the quality standards outlined in ISO-9000 and the certification process has been adopted by the European Community (EC) government.

Many companies obtain certification of compliance to the ISO-9000 standards as a result of pressure from their customers. In the EC, certification is a requirement for firms in particular markets, some of which include: gas appliances, commercial scales, construction products, industrial safety equipment, medical devices, and telecommunications terminal equipment. Certification of compliance is becoming (and in many cases has already been mandated) necessary for countries outside of the EC to supply to European businesses and/or conduct business in the EC.

14.2   The Standards of the ISO-9000 Series

The ISO-9000 series consists of five standards. ISO-9000 and 9004 are guidance standards that are descriptive in nature, and a company would not become "registered" to either of these standards. ISO-9001, 9002, and 9003 are conformance standards that are models for a quality system. They have minor differences depending on whether a facility is involved in design, development, and/or servicing, in addition to production and installation. It is to one of these three standards that a company would seek registration.

The following describes two guidance standards and three conformance standards of the ISO-9000 series that can be applied to various firms:

14.2.1   Guidance Standards

ISO-9000:   Quality management and quality assurance standards— guidance for selection and use. This standard explains the basic quality concepts and defines some key terms. ISO-9000 is used for selecting the proper conformance standard for external quality assurance (QA) purposes and can be used as a guide for the entire ISO-9000 series.

ISO-9004:   Quality management and quality system elements guidelines. This standard is used to give direction for internal management purposes. It can be used by a company to determine which quality system is best for them. ISO-9004 explains most of the elements included in the conformance standards in greater detail and may help one understand the intent of each element.

14.2.2   Conformance Standards

ISO-9001:   Quality systems model for facilities involved in design and development, production, installation, and servicing. There are 20 elements in this standard at the time of this writing.

ISO-9002:   Quality systems model for facilities involved in production and installation, but who are not involved in design aspects. There are 19 of the 20 elements found in ISO-9001 in this standard. Companies that use this model are typically provided product designs/specifications from their customers.

ISO-9003:   Quality systems model for facilities that are involved in performing final inspections and testing. There are 12 of the 20 elements found in ISO-9001 in this standard. Distributors and service organizations would usually seek compliance to this standard.

The 20 elements of ISO-9001 cover the following areas:

1. Management responsibility,
2. quality aystem,
3. contract review,
4. design control,
5. document control,
6. purchasing,
7. purchaser supplied product,
8. product identification and traceability,
9. process control,
10. inspection and testing,
11. inspection, measuring, and test equipment,
12. inspection and test status,
13. control of nonconforming product,
14. corrective action,
15. handling, storage, packaging, and delivery,
16. quality records,
17. internal quality audits,
18. training,
19. servicing, and
20. statistical techniques.

NOTE: At the time of this writing there are potential revisions in process of the element descriptions listed above.

Adhering to many of the above requirements is documentation intensive. There must be proof that a facility knows the requirements, documents procedures that support them, and follows the procedures. Many procedures must be treated as if they were a blueprint, with authorization signatures and release/revision dates. They should be closed-loop, describing what processes precede following a particular procedure to those dependent on the successful compliance to it. In other words, every process within an organization must be outlined and the reader of one should know "from where they're coming and to where they're going." Organization and cross-referencing of documentation and records is very important. And even though the requirements can apply to any commodity or service, there are no "loop-holes" in ISO-9000. A company that is committed in becoming certified, must be committed in modeling their quality systems accordingly.

14.3   The ISO Registration and Certification Process

Certification must be assessed by facility. If a company consists of more than one facility, each must be assessed independently and according to the ISO standard which best fits its capabilities. Once a facility has fully implemented an ISO-9000 standard (i.e., ISO-9001, ISO-9002, etc.) and has documented it with the necessary procedures and policies, it should function in compliance for at least 3 to 4 months before seeking to become certified.

Certification is obtained through a registrar that is accredited to show that it is competent to perform audits. Accreditation varies depending on global location. In Europe, there are two main accreditation organizations and they are government endorsed: In the United Kingdom is the National Accreditation Council for Certification Bodies (NAACB), and in the Netherlands there is the Raad voor de Certificatie (RvC). They certify organizations to conduct third-party quality system audits.

In the United States a registrar can become accredited by the Registrar Accreditation Board (RAB), which is sponsored by the nongovernmentally regulated American Society of Quality Control (ASQC). RAB accredits registrars to certify facilities to ASQC/ANSI QS9000, which is the American equivalent of ISO-9000. While U.S. companies can find that certification to QS9000 is definitely beneficial, it is important to note here that at the time of this writing, this certification may not be accepted in the EC as a valid ISO-9000 certification. If the United States government someday endorsed/regulated QS9000 and certification, it could impact the potential of its acceptance in Europe.

14.3.1   How to Find and Choose a Registrar

One of the most logical ways to choose a registrar is to informally survey your customers for their recommendations by asking them which registrar they would recognize as valid. A decision could then be made based on compiled survey information and contact made with the registrar chosen. One should also be able to contact quality or plastic related organizations or societies that can provide them "leads" or recommendations in choosing a registrar. There is also a U.S. National ISO-9000 Support Group located in Grand Rapids, Michigan, that provides information via computer bulletin boards and materials.

Once a registrar has been contacted, you should make sure that they are accredited through an accepted organization like the RAB (U.S.). Also ensure that the registrar you have chosen certifies facilities that are in the same type of business you are. Plastics-related manufacturers, processors, etc., should be certified by a registrar that knows plastics. If you are a supplier to the automotive industry, they should also be aware of specific requirements that the automobile makers would expect a company to meet.

After deciding upon a registrar, the next step is to schedule a pre-assessment. The pre-assessment will most likely be slotted for months before the actual date of the registration audit due to the amount of registrars as compared to the amount of facilities wishing to become certified. Remember that the assessor works for you. If there are conflicts for any reason with a particular assessor, request a different one from the registrar.

14.3.2   Costs

Many facilities/companies hire consultants to prepare them for certification. The cost of these consultants varies depending on the size of the company and the amount of time that will be needed to make the necessary changes and implement the processes. A consultant could cost between $20,000.00 and $150,000 in the U.S.

The initial assessment and certification process by a registrar can also vary in costs depending on the size of a company and the amount of facilities it seeks to have certified. These costs can range anywhere from $6,000.00 to $50,000.00 per facility. Also beware of hidden costs such as transportation, hotel, and meals. The total fee is typically nonrefundable and does not guarantee certification as a result of the assessment.

After certification is obtained, spontaneous or scheduled "surveillance" assessments will occur every 6–12 months to assure that compliance to standards is maintained. Maintenance of certification through the registrar usually requires additional annual fees (usually around $2,000.00) to be paid for three years, at which time another full assessment (at full cost) would be needed.

15 Systems Implementation and Maintenance

15.1 Top Management's Role

Top management's role in systems implementation is leadership, support, evaluation, and leadership (refer to Exhibit 43). For a system to be implemented successfully, top management should:

1. Provide direction by establishing or agreeing on what the system must accomplish,
2. support implementation of the system by providing all of the means, tools, and empowering responsible individuals to operate the system, and
3. evaluate the flow and usefulness of the system and continuously act on any knowledge gained to improve the system.

To insure maintenance of an implemented system, the same cycle should occur beginning with "Evaluate."

15.1.1 How a System Fails

Any system or program that is truly important to top management should not fail. If a system fails, it is likely because top management was not fully committed to its success. The following are examples of how upper management personnel can allow a system or program to decline or falter:

- The need or intended process for the system was not properly or fully communicated to all who could affect the system's success.
- The individuals responsible for carrying out certain procedures vital for the success of the system were not supported or held

responsible by their leader. Top management did not truly empower those individuals to carry out their responsibilities effectively.

- Top management did not provide the necessary tools, dollars, manpower, training, etc., necessary for successful system implementation or maintenance.
- Top management "washed their hands" of the system or program after initial efforts allowed it to be implemented. There was no further evaluation, leadership, or support offered after implementation.

15.1.2   Top Management Defined

A "top" manager or director is someone in the company who is totally empowered by the owner or president to make decisions and carry out actions. This type of manager is able to say "I'm going to do this" and do it. It is those with this level of empowerment that I refer to as "Top Management." I find it necessary to define this because being given a title of "top manager" does not necessarily mean that one is truly empowered to make an independent decision and carry it to fruition. If the "Top Managers" in a company cannot make a decision without the Owner, President, General Manager, etc., having to approve that decision, then it becomes the role of that overpowering source to hold all responsibility for the successful implementation and maintenance of all systems.

15.2   Plans and Goals

Usually, the method of planning and goals development is incorporated into a companywide business planning process. Rather than discuss the details of related procedures and components of a business plan, the focus here will be on philosophies involved in enhancing the overall planning process. It is these "enhancements" that will aid your company in utilizing its entire work force to develop worthwhile plans that will produce results.

15.2.1   Identify Strengths and Weaknesses

Before individual departmental plans are discussed, most companies know that the process of defining internal weaknesses and strengths of each department is essential. In many cases, there is a forum of upper management personnel that (amongst themselves) embrace this task. I have found that going first to all of the members of a department to beseech their sense of what is strong/weak about their area is extremely enlightening. This can be accomplished through written surveys or by holding an individual departmental meeting initiated by the manager but championed by another chosen employee. The Manager should not stay for the meeting if maximum effectiveness is to come of it because it can be difficult for Managers to keep their opinions silent. If they voice their opinions (at this point), they can substantially affect the input of others. The Manager should meet at a later time to discuss the results of the department meeting, obtain more details, and address the findings with their input.

The method discussed here should result in upper management learning that the employees indeed know their department's weaknesses and strengths with much more detailed information than an "outsider" or even their manager. This method is so important because it gives the employees *ownership in the planning process from the onset.*

The same meeting championed by the employees can be used to solicit what they perceive as weaknesses/strengths in other areas. Once all information is obtained, the Manager will be a more knowledgeable and better participant in the "forum of upper management."

15.2.2   Continued Employee Ownership of Goals

The same process described above to identify strengths and weaknesses could be used as the preliminary step to plan and establish goals. Improvement planning should not be limited to building strategies to correct weaknesses, but to improve/maintain strengths and other areas not addressed through the strengths/weaknesses evaluation. As well, other external audit results (competitive, governmental, etc.) should be considered in determining goals. After this message is relayed to the employees of the department and the final (management forum input) strengths/weaknesses list is complete, goals and strategies can be developed by the team. Again, a team captain/champion should be chosen (preferably by the other employees) who will facilitate the meeting and record information. After the employees have completed goals and strategies, the Manager should meet with them for a review and make any changes/additions that he/she feels are important. During this phase, the Manager should attempt to have the employees agree to their suggested changes/additions. This is important to maintain the employee's ownership in reaching goals by following the established strategies.

The departmental goals and strategies should then be subjected to the company's business planning forum's input. If further additions/changes are suggested, they again should be presented to the employees for their agreement.

15.2.3   Employee Ownership

The main benefit to be derived from practicing goals and plans development as I have described, is employee ownership. Why should I strive to reach a goal that is not mine? If I have contributed to the goal planning process and the development of strategies, the goals are mine and I will personally take interest and exert effort in accomplishing them. The same principles can be applied to any type of planning process.

15.3   Education and Training

Education and training are tools that everyone needs to perform their job. The introduction of a new procedure, system, or program should be implemented after the "tool" of training has been issued to all who can affect the success of it. No matter how well or detailed a procedure is written, it should be formally reviewed with all affected personnel in the form of training.

15.3.1   Review When Training is not Needed

Even if the new system/procedure does not require that any responsibilities receive specialized education or training beyond what they already have, the following should be addressed when introducing it (reference Exhibit 44 for procedure review steps):

1. *Purpose of the new procedure or system*:   If the purpose is not understood by all "players," the process itself may have no meaning to them and can eventually "fall by the wayside."
2. *Responsibilities*:   The individuals (not department) responsible for procedures should be identified and understand their duties.
3. *Resources and tools*:   All "players" should have a clear understanding of what tools (documents, equipment, etc.) will be required for them to properly perform their part of the procedure. Resources (reference sheets, people, etc.) should also be identified for them, including all responsible individuals for the development of the new procedure. Once the new process is implemented, they will know where to turn if questions or problems arise.
4. *The step-by-step process*:   The flow of the process from beginning to end should be reviewed with all participants.
5. *Questions and answers*:   It is much more effective to address typical questions or concerns with a new system or procedure at the onset. It is an opportunity to make final changes to the new process before implementation and for all involved personnel to share their ideas, concerns, and questions.

15.3.2   Education

Formal education is required when knowledge in a particular discipline, tool, or area is important to put a system or procedure together. Specialized education is also important for any individuals who need it in order to effectively play their role in a process.

Today, a large group of colleges and universities offer education in Statistical Process Control, Total Quality Management, Plastics, and many other quality-related tools and topics. In addition to the universities (reference Exhibit 45), the following sources can be sought out to provide specialized education:

- *Customers* in many cases can provide your company with training in certain areas or refer you to another source to obtain the education.
- *Chambers of Commerce* at both state and local levels provide education (usually free) in the areas where its members request it. It is possible that quality related seminars are already on their

agenda. If not, contact them and inform them of your company's needs.

- *Professional organizations* that your company or its employees belong to are an excellent source of education resources. Aside from quality- and plastic-related seminars they may hold within the organization are other sources for education that they will be happy to provide you with.
- *Mailings* from various training and education organizations and advertisements for seminars float across the desks of everyone. Send out a memo to all personnel requesting that all related mailings of this type be forwarded to the training and education coordinator. The coordinator can maintain a "library" of education sources for future reference.

15.3.3   Training the Trainer

If the intention of sending an individual for formal education is to have him/her gain the knowledge required to train other personnel, evaluate the individual's ability to train first! There is a distinct difference between understanding a topic or discipline and teaching it to someone else. The ability to learn does not automatically give someone the ability to teach. It is very possible to have someone of average intelligence turn out to be a better teacher/trainer than a person of very high intelligence. The best trainer is not the "learner" but the "interpreter." The following characteristics should be assessed when looking for a person who is to potentially train others:

1. Is the person charismatic? Do they keep your attention and are they easy to listen to and understand?
2. Can the person effectively develop analogies and examples that make it easier to understand what they are attempting to convey?
3. Does the person take advantage of visual aids and workshop training methods that provide "hands on" training and/or role playing?
4. Is the person's verbal tone, flow, speed, and enunciation easy to follow and understand?
5. Are there any quirks, habits, or types of gesturing that may distract an audience?
6. Is the person adept at modifying his/her training methods and

vocabulary to easily relate to persons of significantly less intelligence? . . . more intelligence?

15.3.4   Evaluate the Effectiveness of the Training

Whether training is being conducted in a certain discipline, to increase overall knowledge or to implement a new system/procedure, the effectiveness of that training should be evaluated. The best way to evaluate the usefulness of the training is to issue a questionnaire at the end of the session to all of the attendees. I personally do not recommend tests that can be scored or indicate whether or not an individual "passed" or "failed." Tests of this type can create a tense atmosphere and focus on the effectiveness of an individual's ability to learn rather than the ability of the instructor to teach.

Questionnaires can be set up like tests. It should be made clear to the employees that the only purpose is to evaluate the effectiveness of the training only. Stress to them that no one will "pass" or "fail" the session.

The results can be evaluated to identify areas where more time should have been spent, visual aids used, or participation encouraged. If there is a small percent of personnel that did not grasp all that they were intended to learn, it's probably that extra one-on-one training will be required for those individual(s).

15.3.5   Ongoing Training

A training plan should be developed at least annually that includes "refresher" sessions in the use of certain tools, disciplines and procedures. For example, SPC, use of measuring equipment, and other types of training should occur throughout every year for affected associates. With employee turnover, day-to-day repetition, organizational changes, etc., ongoing training in particular areas is crucial in maintaining established systems.

15.4   Follow-Up (The Art)

Effective "follow-up" is a key element in maintaining any implemented system (new or otherwise). Some of the types of follow-up that have already been mentioned include assessments, audits, and validations. But just as important, informal follow-up can many times be more timely, objective, and "unrehearsed" to the point of providing insight that could be overlooked during audits.

Follow-up lets everyone involved in the process understand that management is indeed very serious about its success. It also reminds them that the role they play in its success is consistently being assessed by management. Follow-up provides an avenue for continuous improvement during both implementation and maintenance stages of all systems, procedures, programs, and processes.

Follow-up should be integrated into management's daily planning process and should be consistently taught to all leaders throughout the company.

15.4.1   Becoming Disciplined in Daily Follow-Up Planning

Most good leaders use some source of time management system (Franklin Planner™, Day-Timer™, etc.). The art of follow-up is in integrating it into a time management system. For any system, procedure, process, etc., that you or one of your employees champion, are responsible for, or have ownership in, following-up on its progress is essential in maintaining it. Practice these steps:

1. Remind yourself to follow up,
2. follow up,
3. follow up on the follow-up,
4. remind others to remind themselves to follow up, and
5. continue follow up as necessary.

For example, if a "Job Launch" procedure has been developed, just implemented, and it is one of your employee's responsibility to oversee the process, the following practices of follow-up might be employed:

1. Advance one week into the future of your time management system (calendar) and write "follow up on job launch program".

2. In one week, your calendar would remind you to "follow up on job launch program." You would then contact your employee responsible and inquire:

    How is the new job launch program going?

    Are we up-to-date according to the schedule? and

    Please give me a list of upcoming job launches and the expected dates of completion by tomorrow afternoon.

    This whole process will consume approximately three to five minutes of your time.

3. Immediately after this conversation, mark on tomorrow's calendar to "receive job launch list from _____ by afternoon." If the list is not received by the end of the day, follow up.

4. When you receive the list, review it with the responsible employee. Instruct them to log each one into their calendar two (2) days before it is due to prompt themselves to remind all other persons involved in the launch of due dates. You will have succeeded now in "practicing what you preach" by instructing someone to follow up by following up.

5. Now you will advance maybe two weeks into your calendar and again remind yourself to "follow up on new job launches" and "make sure _____ is properly following up". As you continue this cycle of follow-up, the frequency will become less often as you assure that your employee is properly following up. Eventually, you may remind yourself every couple of months to inquire as to how well the program is going. Please reference Exhibit 46 for a visual aid of this process.

It is wise to never consider anything "completed" just because it is "started." Continue advancing into your calendar to remind yourself to follow up rather than "check it off" as being completed. When comments are made like "He never forgets anything" or "She's always on top of things," you will know you have been effective at the art of following up.

Exhibit 43: Top management's role (systems implementation)

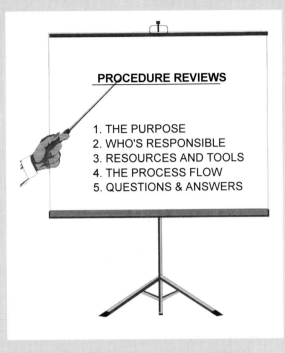

*Exhibit 44: Procedure review steps
when new training is not required*

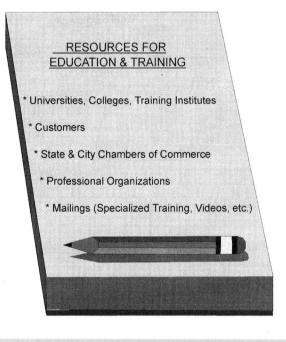

Exhibit 45: Employee education alternatives

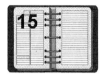

1. Remind yourself to follow up

2. Follow up

3. Follow up on follow up

= *Effective Implementation
*Maintenance
*Continuous Improvement

4. Remind others to follow up

5. Continue to follow up

Exhibit 46: The art to follow up

16 Continuous Improvement

16.1 Identifying, Tracking, and Reporting Indicators

For every area of focus in each department, the effectiveness of continuous improvement efforts should be measured by tracking certain indicators. An indicator is something that can be measured. It will provide critical information as to the success of continuous improvement efforts. It also can be charted and evaluated at established frequencies so that trends, freaks, and other patterns can be identified and necessary actions taken.

16.1.1 Identifying Measurable Indicators

Identify what can be measured that will indicate whether or not a particular area, goal, or activity is performing as expected. There are usually many indicators that can be identified to measure performance in an area of focus. Parts per million (PPM) is a very popular indicator in quality that is a calculation of parts shipped vs. parts unacceptable. A PPM of 25 would mean that for every million parts shipped, 25 were rejected. But there are many other creative indicators: for example, if one of the Quality Department's goals was to improve the efficiency and effectiveness of internal problem solving reports, the following indicators are a few that might be identified as measurables:

- Average time turnaround for a problem solving report to be completed,
- number of reports opened each month that are recurrences, or
- average percent of opened reports that are completed within a week.

One or more measurables should be chosen to indicate performance in focus areas. The number of measurables and the type of measurable may change throughout the tracking process for the purpose of a shifted focus, problem solving, or overall tracking improvements.

Indicators may be "broken down" to provide more information for problem solving efforts. For example, if the measurement indicator of "Number of recurrences" was being monitored and results showed an upward trend or stabilization, the chart can be "broken down" to provide more information. Some "breakdown" charts that could be added as measurement indicators might be: "Number of recurrences for each product type", "Number of recurrences for each material category," etc. Once control is regained for the main measurement indicator, it may not be necessary to continue compiling information for and reviewing any broken down graphs.

16.1.2   Tracking Indicators

Once a measurement indicator is determined, the method of compiling the information for tracking purposes should be established. Tracking would also require that an individual be assigned responsibility for the task. In the above examples of measurement indicators for the improvement of "efficiency and effectiveness of internal problem solving reports," the following methods of compiling information and tracking might be used:

- For "Average time turnaround for a problem solving report to be completed": The Quality Manager will review the "opened" and "closed" dates on all completed reports. He/she will record the number of work days it took for each report to be completed. He/she will then calculate the average number of days to complete a report and chart that figure on a line graph monthly.
- For "Percent of reports opened each month that are recurrences": The Quality Manager will investigate each opened report and determine if the concern is a recurrence by checking the tracking log back one (1) year. He/she will keep a tally sheet to record the total number of reports opened and the number of recurring concerns. He/she will calculate the percent of recurrences and record it on a bar graph monthly.
- For "Average percent of opened reports that are completed within a week": The Quality Supervisor will review each completed report and keep a tally sheet to indicate the number of reports that were closed within a five (5) work day period. He/she will calculate (monthly) the percent of reports closed within a five (5) day period

based on the total amount of reports opened within that month. A bar graph will be used to track the monthly results. Refer to Exhibits 47, 48, and 49 for examples of tracking charts and "broken down" charts.

16.1.3   Reporting and Reacting to Indicators

To track indicators is a futile task if there is no formal review to report and react to results. The purpose of tracking indicators is to share the data at an established frequency with a multidisciplined team to accomplish the following:

1. Review the purpose for the indicator and expectations/goals.
2. Evaluate the chart (as a team) to determine:
   a. If expectations/goals are being met,
   b. if there is a trend, freak, cycle, etc.,
   c. if past actions improved behavior of the indicator,
   d. what the root causes are for the behavior of the indicator,
   e. what actions can be taken to improve the results or move closer to the expected target, and
   f. whether further information is needed to better evaluate root causes and take action.
3. Record monthly results, observations, root causes, and actions into a monthly meeting report.

I suggest that a monthly meeting be used to review indicator results from all areas within your company. Invite employees from all levels of the organization as they can provide valuable insight during discussion. It is also an avenue to get word back to the shop floor that the company is very serious about continuous improvement in all areas. As your company continues this process, graphs that show monthly results during a year can be compiled to show year-by-year results. Yearly and monthly graphs can be posted in the plant for all employees to review.

16.1.4   Helpful Hints

Someone within the organization should be given the responsibility of maintaining graph history for all areas and indicators. The same person should be responsible for monthly reports and annual reports. They should also bring all history to the monthly meeting so that meaningful observations can be made and questions asked.

Charts should depict the results of the indicator for each month. For example, a bar graph on "Prevention vs. Detection" within the "Cost of Quality" report, should show two bars: one indicating "prevention" dollars and the other "detection" dollars for January, two bars for February, two for March, etc. To simply look at the results for the past month will not enable the team to evaluate trends, associate history, etc. Once data has been collected for a year, keep twelve (12) months on the graph. In other words, if the report month is April, the chart would show indicator results for May of the previous year through April of the current year. Also, always show the goal line on the chart.

16.2   Eliminating Gray Areas

"Gray areas" are not "black and white." They are areas of question that result in subjective decision making in inspection and at the manufacturing process.

The continuous elimination of "gray areas" in quality standards will improve the overall decision making process, resulting in more consistent and improved product quality. Every year, special emphasis in specific areas should be documented in the form of a plan to eliminate gray areas. An alternative to an annual plan would be an ongoing employee involvement group that continuously tackles the problem of gray areas.

16.2.1   Examples of Gray Areas

The following are three examples of common gray areas that cause inconsistent decision making in regard to product quality:

16.2.1.1   Good/Bad Visual Aids with Windows

Visual aids are essential on the manufacturing floor to make decisions in regard to the aesthetic quality of products. A mistake that is often made is the posting of an "acceptable" product and "unacceptable" product that leaves a gap or "window" (or gray area) between the two. If, for instance, a small burn on the back of the product is allowable, there should be "good" and "bad" visual aids that are very similar. One would show what is acceptable and the other what is not acceptable. The operator should be able to determine how much of a burn is too much. In many cases, a visual aid board will exhibit an acceptable product with a tiny burn and an unacceptable product with a large burn. When the process makes a product with a burn whose size is in between the samples (in a gray area), a subjective decision must be made as to whether the part is acceptable or not. The next operator may very well question a part with the same burn and make the opposite decision as to whether the part is acceptable or not. Continue closing the window between examples of acceptable visual aids and those that are unacceptable.

16.2.1.2   Inspection Criteria with Ambiguous Wording

Inspection instructions and operator instructions should contain no gray areas if they are to be understood the same by all employees who are expected to follow them. The following are some examples of how to identify and eliminate gray areas in inspection criteria:

Gray Areas	*Black and White*
Slight parting line flash is acceptable	*No more than 0.5 mm* of parting line flash is allowable
Small burn on one of the back ribs is acceptable	*Refer to visual aid sample* board for *maximum* burn that is acceptable on *any* rib on the back of the part.

How much is "slight"? How small is "small"? Are burns acceptable if more than "one" rib is affected? It is these types of questions that are asked by inspectors, operators, and foremen when instructions are not worded so that decisions can easily be made. It is also these questions that will inevitably result in the opposite decision being made on the next shift for the same discrepancy. Make instructions explicit and whenever a question or concern arises as a result of interpretation, improve the instructions immediately.

16.2.1.3   Instructions and Procedures That are Not Enforced

Out of all gray areas existing within a company's instructions and criteria for decision making, this one is the most detrimental. It is the most damaging to good decision making practices because it creates gray areas in already effective and explicit instructions. Even if all visual and written tools that aid in consistent decision making were "black and white", the act of them not being enforced will inevitably result in chaos. If inspection instructions state that the "hole size must be 23.0 +/- 0.5 mm," then a hole size larger or smaller than that specification should be rejected. If a manager tells an inspector or operator "well, it's just 0.2 mm above spec., we can let it go," the manager has just eliminated the hole size requirement. The next time the hole size is 0.3 mm above specification, the Inspector may rationalize that the hole size really doesn't matter or that it is just .1 mm larger than the last time. In addition to this problem comes the fact that an explicit written instruction did not have to be followed. It stands to reason that other written instructions do not have to be followed. *If written or visual instructions must be deviated from, change them to match what is being considered acceptable.* Formal written deviation approvals are another method to justify nonconformance to standards, but they are not as effective as creating instructions that can be followed. Regardless of the fact that a deviation approval specifies that only a certain quantity of product is affected, the Inspector and Foreman *will remember and believe "if' it's OK for those, why not these?"*

16.2.2   Make New Decisions Only Once

While it is possible to eliminate and not create gray areas in product quality decision making, it is impossible to reach a point where gray will simply stop presenting itself. With the inherent variation in processes, materials, etc., and the never-ending flow of the introduction of new jobs, questions will arise. A good philosophy is to never let the same question be asked twice. As questions arise in regard to the acceptability of product quality, improve visual and written instructions immediately so that the same judgement will be made again next time.

16.3   Employee Involvement

Involving employees in continuous improvement efforts creates a definite "win/win" situation. Employees become empowered in the decision making process, gain ownership in results, and achieve self-esteem by being included in the information flow. The company gains valuable insight by utilizing input from the experts on the manufacturing floor. Involving employees to continuously improve all areas can range from simply including them in meetings and posting information for feedback to establishing full blown employee involvement groups (EIGs) to tackle certain concerns and issues.

16.3.1   Sharing and Soliciting Information

This form of employee involvement opens the lines of communication between the employees and management/supervision. Often management is unaware or has forgotten exactly what it is like out "in the trenches." The majority of manufacturing employees can be blind in regard to management's concerns and efforts. Simply opening the lines of communication by involving each other creates a synergy that alleviates tension and fosters cooperation throughout the company. Everyone's level of knowledge is important and should be used to continuously improve processes and systems.

It is management's responsibility to create an open environment of two-way communication. The following suggestions are examples of how to begin involving employees (reference Exhibit 50):

- Post information for employees that will show them what areas they can influence, where the company wants to improve, and good things (i.e., training) that the company does for them. Posting cost information will help employees realize that not only does that cost exist, but it is significant (e.g., rework costs, premium freight cost, returned goods cost, training costs, absenteeism, etc).

- Invite floor employees (at random) to visit or "sit in" on many of the ongoing meetings that your company holds, (i.e., Quality Circle Meeting, Engineering Meeting, Production Meetings, etc.). The employees will learn how much time, effort, and planning that the salaried employees really endure each day. They will understand the importance of the various positions within the company and will share that with their co-workers.

- Implement and *maintain* an employee suggestion box program. Make sure that all employees are aware of the communication tool. Check the box *daily* and *react* to all comments, questions, and ideas in writing or in person. The Personnel Director could head up this program and disseminate all suggestions to the appropriate Manager.
- Use a company newsletter to convey company *news* as well as informative articles and personnel related news. Some examples of company news might include: upcoming customer assessments, expansion efforts, expected dates of implementation of new programs and benefits, articles that describe the process and reasoning involved in bringing a new procedure, system, etc., to initiation.
- Encourage a company-wide (ongoing) campaign to get all managers out onto the manufacturing floor, (this is still *employee involvement*). All managers in every area could familiarize themselves with employee names, know who the "natural leaders" are, and understand the manufacturing processes. Give an employee a five-minute break once a week by taking over their job. It will more than likely be the best five minute investment you will have made all week towards "opening the lines of communication."

It is important to alleviate competition as much as possible when posting informative charts, graphs, memos, etc., for employees. Try not to break down costs by shifts or areas. There is always a root cause (or excuse) for the major contribution of undesirable results. People feel "blamed" if the approach in providing information points a finger at them. Post companywide results. If a particular shift is the main influence on an undesirable result, create the chart/graph breaking the information down into shifts, but use it only to work with the responsible shift to solve the problem.

16.3.2   Formal EIGs

Employee involvement groups (sometimes called teams, task forces, etc.) can be productive in: meeting goals, establishing systems and procedures, solving and preventing concerns/problems, and in improving systems, processes, programs, etc. Basically, the establishment of an EIG will aid in any circumstance where a need is recognized.

Once a need is recognized, steps similar to the following should occur (refer also to Exhibit 51):

1. *Form a team*:   Discuss the idea to form an EIG with co-workers and other key persons who might benefit the team to get an idea of who to invite into the group. Make a list of potential candidates and contact each employee. Discuss the "need" with them and the decision to use an EIG. Request that they be involved and ask what is the best time for them to meet. You should not *require* that an employee join an EIG! Schedule the first meeting, issue a meeting notice, and chair the first meeting.

2. *Meet and issue minutes*:   Discuss the need in detail and involve the team in brainstorming ideas to satisfy the "need." Have all members agree on who should be the team captain and conduct a vote. Discuss and decide on the frequency of the meetings and establish the day and time for all future meetings. Issue minutes for every meeting.

3. *Highlight objectives*:   Decide from the brainstorming process (as a team), what the team's objectives are. Document the objectives.

4. *Create strategies*:   Review the documented objectives and develop strategies to meet them. Always assign responsibility and establish expected completion dates.

5. *Measure success and report*:   Decide on measurement indicators (refer to section 16.1, "Identifying, Tracking, And Reporting Indicators" in this chapter) for each objective.

6. *Satisfy the need*:   The need should be satisfied when conducted strategies have met the team's objectives.

7. *Report results and congratulate the team*:   Report all results and congratulate all team members.

Special recognition could be given to all formal EIG members who belong to a successful team (satisfied the need). I strongly recommend non-monetary recognition because these types of "bonuses" will inevitably turn into a perception by many employees of the company "taking money from" certain employees. Recognition could be in the form of a certificate of achievement, names and accomplishments in the company newsletter, verbal congratulations during a company meeting, etc.

One should never consider an EIG to be "finished." It can serve as a permanent resource for the type of problem it solved. Keep a list of all EIG's initiated and their members. After the EIG has met all established objectives, it should be considered "inactive" unless the same or similar need arises in the future (at which time the team would become "active" again).

16.4   Employee Incentives

Motivated employees play a key role in assuring continuous improvement in any organization. While there has been much controversy as to the best way to encourage employees, I believe all methods involve some form of recognition. Acknowledgment is the first reason any of us initially shared our special abilities. As infants we cried for recognition, as children we proudly recited the "A, B, C's" for recognition, and as teenagers we may have done a chore (without being told to) for recognition. Being appreciated helps us define our own meaning of value.

16.4.1   Compensation Motivates an Employee to Show Up to Work

Compensation may be an incentive to take a job or a promotion but it soon becomes what the company owes the employee for services rendered. It may keep employees motivated to show up to work every day and perform the duties required to keep their job, but the fact is, you cannot *pay* someone to be truly motivated beyond that. Whether your company pays an Operator $5.00 an hour or $20.00 hour, the employee's lifestyle will adapt to that compensation. And employees will most likely always feel that their time is worth double what they are paid. Even if an employee thinks he/she is over-compensated, the stress involved in trying to perform to that level of pay will often overwhelm the effectiveness of that motivation.

Multi-tiered pay scales or wage increases based on individual job performance motivates only a small percent of a company's personnel base. It may create a much larger percent of poor attitudes that are not motivated and may even cause sabotage. Job performance evaluations are subjective. What one supervisor considers to be excellent job performance is what another perceives as average job performance. Employees who are at the low end of a pay scale or increase will justify (to themselves and to others) how the company was unfair because. Another perception of tier wage and increase structures may be that *the company will accept inferior job performance*. Less is to be expected of an employee who is paid less to perform the same job as his/her higher paid co-worker. A consistent pay structure for job classifications that are identical can send a clear message to the employees that the company expects the same level of job performance from each of them.

16.4.2   Monetary Bonuses are Short-Lived Incentives

Monetary bonuses for certain achievements soon turns into the company *taking money* from the employees. A bonus starts out as a "good job" gift and may initially motivate employees. The bonus soon becomes *expected* and the first time it is not received will reap employees yelling of unfairness, accusations of company manipulation of statistics, and "finger pointing" to place the blame anywhere but on themselves. The bonus will no longer be a "gift" but a *penalty*. These repercussions can definitely demotivate employees and cause the bonus incentive program to lose its affect.

If, for example, a bonus program was initiated to reduce customer returns for poor product quality, certain guidelines would be developed. Let's assume that a $50.00 bonus will be issued each month to each employee on a shift that has no customer returns during the month. Initially, this incentive will be very motivating to employees. They will rally together to achieve their goal. After the first month, maybe the employees on the afternoon shift will each receive a $50.00 bonus. That shift is still motivated, as well as the others who can see that the bonus is achievable. However, some finger pointing has already began on the other shifts because those employees want to solve the problem of them not receiving the bonus. After a couple of more months, one of two things is going to happen:

1. The employees who have been able to achieve the goal will begin to *expect* and count on the bonus every month. The first time they don't get it, hostility will surface, bills will not be able to be paid, certain employees who are deemed responsible will be harassed, and the bonus is no longer a "good job" gift but a *"bad* job" penalty.
2. The employees who have not been able to achieve the goal will begin to accept the fact that it is unattainable. They will think that packing 100% acceptable product is out of their control and any motivation derived from the bonus program will disappear.

Even to develop a bonus program based on individual efforts can "backfire" for the same reasons that tiered compensation structures do. A company has to meet objectives as a whole to succeed and improve, not by a few individuals' accomplishments. The fact is, those few individual accomplishments will probably be evident with or without a bonus incentive program.

16.4.3   Back to Recognition

Many managers and supervisors find no reason to thank employees or "pat them on the back" for accomplishing what they are expected and paid to do. Employees are not paid to make mistakes, have bad attitudes, or slack on performing their duties. Therefore, when an employee slacks, makes a mistake, etc., they get the full attention of their leader during counsel (some call it "being reprimanded"). Acknowledgment is important to everyone. Some people don't "follow the rules" for the sole purpose of being acknowledged. Even children will go to the extreme of soliciting punishment as a form of getting attention from their parents if they are not getting it otherwise. This message is not to suggest that one should not acknowledge poor behavior by their employees but rather consistently recognize and pay attention to employees whose behavior meets or exceeds their expectations. This practice will encourage those employees to maintain that level of attitude and performance and will even motivate them to further improve.

Money would be better spent to educate and train all managerial and supervisory personnel on how to motivate their employees than to facilitate bonus programs that will probably fail. Some of the following are examples of recognition that will motivate employees to accomplish what they are capable of:

- A manager spontaneously pops their head into the office of an employee and says, "By the way, just wanted to let you know you're doing a great job. You've been handling things so well on your own lately and I want you to know I appreciate it."
- A supervisor in manufacturing walks over to an operator that has had a problem in the past with a poor attitude and says, "Hi! You've been in a pretty good mood the last couple of days. You're turning into a person that I look forward to working with every day. Keep it up—it makes me feel good when employees smile."
- The General Manager holds an awards meeting every year with all employees in attendance and gives a certificate of achievement (mounted on a nice plaque) to employees who have had excellent attendance records, have been with the company for over ten

years, have significantly contributed to major improvements in the company, have excelled beyond normal expectations in their job, or any other reason that an employee may deserve recognition.

- A manager periodically walks through manufacturing and approaches various employees and gives them compliments like, "You're a whiz on this job! I know this is a tough job but you seem to make it look easy", and "Hi! I thought I'd take a walk around today to remind myself what a good job my most important employees do."
- A supervisor takes time out of his/her busy schedule to visit at least one different employee every day and build on that relationship by acknowledging that the employee is a person (not an extension of machinery). He/she may inquire, "How is that son of yours doing in college?", "What did you think of the hockey game last night?", or "Are you still going to those arts and crafts classes?".

Good relationships build loyalties that nourish motivation.

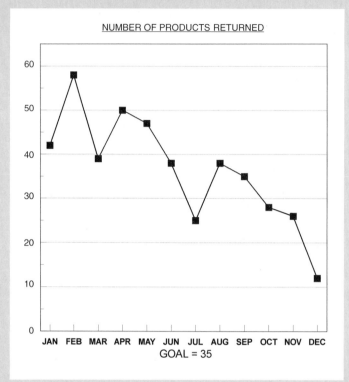

Exhibit 47: Customer returns per month
(tracking chart)

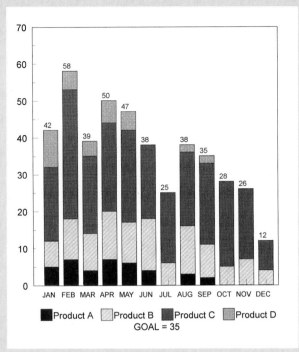

Exhibit 48: Customer returns broken down by product line

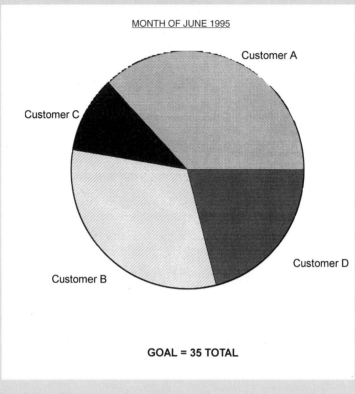

MONTH OF JUNE 1995

Customer A

Customer C

Customer D

Customer B

GOAL = 35 TOTAL

Exhibit 49: Customer returns broken down by customer

* Post Information (i.e., Rework Costs, Absenteeism Rates).

* Invite Employees To Meetings (i.e., Quality & Production).

* Implement & Maintain An Effective Employee Suggestion
  Program (i.e., Employee Suggestion Box).

* Publish A Company Newsletter.

* Get Management Out Onto The Manufacturing Floor -
  There Is A World Of Information Out There.

*Exhibit 50: Examples of how to begin
involving employees*

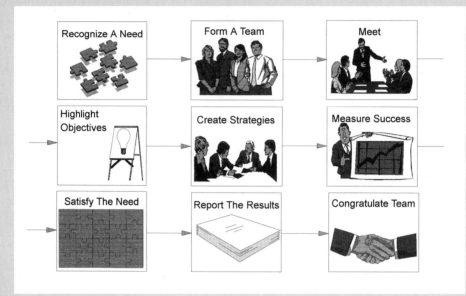

Exhibit 51: EIG process for continuous improvement and problem solving efforts

17   Relationships

Good working relationships are vital to the overall success of any company. The relationships within each department, between departments, and with suppliers and customers are the substance for the flow of all processes. It is the responsibility of the top leaders in a company to define a philosophy on how relationships should be perceived by all employees. It is then their responsibility to encourage that interpretation and behave (as role models) according to the principles they have developed.

17.1   Departmental

The leaders within a department are expected to support and provide direction to their employees. The connection between these two could be likened to a "doctor/patient" relationship. The leader (doctor) is an expert in their area who guides, strengthens, and appreciates their associates. An associate (patient) respects their leader, follows his/her directions, and relies on them for support. The doctor is not the patient's " boss," nor is he/she "over" the patient. The doctor is a leader that *works for the patient*. In fact, if not for patients, a doctor would have no job.

There are many ways the leader/associate relationship can be perceived and vary from the way I choose to interpret it. While I am sure many other analogies could be developed, the above example has been proven with resounding success for myself and many others.

Another relationship within a department is the one between two co-workers. Whether their titles are the same or different, they have the same leader. The way co-workers interact with each other is strongly influenced by the expectations of their leader. The leader should make it clear, verbally and through actions, that they will not abide animosity between co-workers. Co-workers must function as a team to create the synergy necessary for optimum departmental productivity.

If the channels for good communication are open and the leader is actively involved in assessing working relationships within their department, they will know when intervention is needed. They should allow employees adequate time to work out differences on their own. If the employees fail to fully resolve any conflict, the leader must become involved promptly enough so as not to allow the faltering relationship to affect productivity and quality (or other relationships). If the leader has to take action, the approach should be as such that it does not belittle the employees. The leader should first acquire a keen understanding of the problem and convey a sincere need to resolve it. He/she should make sure that the co-workers both assume ownership in any actions that are meant to re-establish a good working relationship. This is accomplished by guiding the employees to actively suggest and agree on solutions themselves.

17.2   Interdepartmental

Interdepartmental relationships are those between the individuals from one department to another. All areas within a manufacturing organization need to bolster each other with the common goal of supporting manufacturing. Even the manufacturing department needs to provide information and feedback to other departments so that they can be better assisted by them.

It is good that people take pride in their line of work and know that the contributions of their department are vital to the overall success of the company. But no department can boast of maximum productivity until they learn the benefits of performing interdependently within their organization.

To have the leaders of each department work interdependently together, it is important that each have a good understanding of their coworker's responsibilities. This insight promotes continued understanding which is essential for a leader to maintain good relationships with other departments. For example, if the performance of one department affects the productivity of another, animosity could build to the point that each department behaves as a separate entity within the company. If the leader of the affected department has a true understanding of the other's responsibilities and their relationship is good, they will be compassionate and helpful rather than angry. Top management needs to assume the majority of ownership in how their leaders understand each other and work together.

Many companies have already began cross-training or "job switching" programs that allow leaders to gain understanding of each others job responsibilities. These practices make a lot of sense in regard to building relationships. The following two scenarios will demonstrate how important *understanding* is in maintaining productive interdepartmental relationships:

17.2.1   Scenario I—Poor Understanding of Other Departments' Responsibilities

Jerry is the Quality Manager, Manny is the Manufacturing Supervisor, and Paula is the Production Control Manager. Jerry's people have rejected stock currently running in manufacturing that is considered unacceptable by the customer. Manny has tried to improve the process all day and cannot believe that QA rejected them for a "tiny little stress mark." Manny thinks Jerry doesn't care at all about how this rejection affects his entire day. He approaches Jerry and says "I can't believe you're rejecting these!" Jerry says, "If you had the process running like it did last run, we wouldn't have to reject anything!"

Paula has received the news and joins into the debate by adding, "I have to ship all of those goods by 4:00 p.m.! The customer should have had them by yesterday!" Manny glares back at Paula and says, "If you needed 'em yesterday, you should've scheduled the job to run last Friday!"

Tempers rise, accusations fly, and either the problem will go unsolved, or top management will have to intervene and give orders. Being subjected to a decision by their manager(s) left at least one of these people feeling that they were not understood. Bad feelings between the three still exist even after the problem has been "solved."

17.2.2   Scenario II—Good Understanding of Other Departments' Responsibilities

Diane is the Quality Manager, Mary is the Manufacturing Supervisor, and Paul is the Production Control Manager. Diane's people informed both her and Mary of defective product on one of the machines. They haven't rejected the production yet because they know Mary will immediately panic. Diane explains to Mary that she knows the customer will reject the goods in their receiving location if she passes them—the customer has rejected the same defect before. Diane knows that if Mary could simply turn a knob and erase the defect, she would have already done it. Mary says, "I just can't figure it out, the last run was fine."

Diane and Mary decide to go see Paul immediately before he overhears their dilemma in a hallway conversation. In the meantime, QA is placing reject tags on the stock and getting it segregated. Diane and Mary inquire to Paul, "How critical is it that you have these parts today?" Paul says, "Well, I was supposed to ship them yesterday. They gave me a late release and I couldn't run them prior because I didn't have any 750-ton press time." Diane then begins to follow up on her department's process of supporting manufacturing. She verifies that the raw material properties are consistent with those from the last run, etc. Mary continues to trouble-shoot the process and looks at the tool.

The three of them have the same goal—get acceptable product to the customer today. They will solve the problem today or contact the customer to apprise them of the situation and get further instruction. The point is that they are all in it together. They understand they all have the same goal and did not blame each other. They will definitely be more productive than Jerry, Manny, and Paula in handling this situation.

17.3   The Customer

The relationship between a company and its customers depends strongly on the individuals within the company who deal directly with the customers. All contact with the customer is critical at any level. Whether dealing with an hourly worker or the president at a customer location, the same level of respect and performance should be displayed. It is expected by them.

Customers are just like you. They depend on the performance of their suppliers. They are sometimes not very knowledgeable about your processes. They have only so much patience. They are trying to perform well in supplying to their customers. They know you are their supplier and frankly do not wish to hear about your other customers (especially if they were forsaken because of them). They are definitely more interested in solutions than excuses when there are problems. You always hear from them when there is problem but very rarely when there is not. Customers are just like you.

Every person within an organization who is to have contact with a customer should have experience and knowledge at doing so. If they do not, they should receive training on how to accommodate them and communicate with them effectively. Any individual who has contact with the customer must realize that he/she represents your company—not just themselves or their department. As a representative, they need to assume complete responsibility for the company when dealing with a customer.

I do not condone yelling and screaming at anyone (including suppliers), but let's face it; it does occur. If, for example, you were at a customer location (or on the phone), and the customer yelled in an angry tone, "I can't believe you shipped this stuff to us again! You shut the line down last month for the same reason! You just don't care, do you? Do you think I complained last month because it's fun? I'm gonna charge you $120.00 hour to sort through these products!"

I suggest the following approach:

DO NOT:

Do not yell back at the customer.

Do not hang up on him/her because you simply cannot tolerate that kind of treatment.

Do not place the blame elsewhere by replying, "I did not ship anything to you. I do care. It's those people out in our manufacturing shop that did it. I'll tell them they better get us out of this mess." The customer will not feel very secure about the problem being solved.

DO:

Do tell the customer that you personally take full responsibility for the problem.

Do tell him/her that you will assure all in-house stock is immediately quarantined and will have Quality re-inspect the product after manufacturing sorts it.

Do tell the customer you'll have a representative at their location immediately to sort stock there.

Do tell him/her, "I apologize for the unfortunate recurrence of a problem that we should have prevented. Not only do we need to correct this problem but we need to correct whatever we failed at last month in our problem solving efforts. I understand how this makes you feel but the fact is we do care and will do whatever is necessary to regain your trust in us."

The customer needs to end the conversation and feel secure in the fact that the person he/she spoke with will take care of their concern. Maybe that customer representative had just been reprimanded by their leader because he/she was not effective in getting you to solve the problem last month. The customer representative that you speak with has to answer to their leader the same way you do. The important thing is to convince the customer that he/she spoke with the right person and continue ownership of the issue until it is prevented.

17.4   The Supplier

Your company's relationship with its suppliers is as important as that with your customers. Your customers rely on your suppliers through your company. Supplier expectations should be communicated with them on a continuous basis. Being proactive by documenting guidelines, frequently communicating issues, providing feedback, and developing the supplier will result in establishing a good working relationship that will benefit you and your supplier.

Behave professionally with suppliers. It is not necessary to post and issue information or charts that depict the "Top 5 Worst Suppliers" to your company. It is not productive to force supplier conformance to your company's requirements by focusing on providing feedback of how "bad" one might be. A good relationship results from effort in developing a supplier that is not adequately performing rather than "hammering" them with threats. Desourcing a supplier should be a *final* resort and should not occur often, if at all.

To assure the practice of cultivating suppliers to perform at the level expected, a process could be documented on "supplier relations" for Purchasing and Quality to refer to. A sense of responsibility for supplier effectiveness should be assumed by your company. Play an active role in working with suppliers to solve problems by taking ownership in preventing any concerns.

A good working relationship with suppliers as a result of:

- Frequent communication,
- providing documented guidelines outlining expectations,
- providing good feedback on performance and for their successful traceability efforts,
- being a partner with them in solving problems,
- developing them, and
- not demeaning/belittling or "hammering" them should result in your company's control of its supply base, and loyal and dedicated suppliers who can meet your expectations.

17.4.1   Little Customer—Big Supplier

A large percent of plastic injection molders, blow molders, plastic extruders, assemblers, plastic die-cutters, etc., are what can be considered as small and medium-sized companies. It is very likely that many of these companies will need (or be subjected to) what is considered a large company as a supplier. Many of the "large company" suppliers will treat your company with the respect that a customer should be treated, (especially if your company is professional, has documented guidelines, and an effective purchasing function).

Unfortunately, regardless of professionalism, outlined expectations, and customer status, it is possible that your company may have no choice in retaining a large supplier that is uncooperative in regard to your expectations. Some of the following are examples of how your company might be subjected to these types of suppliers:

- Your customer requires you to buy from that particular source.
- They are the only supplier of the item you must purchase.
- They were the only supplier who would provide a low volume of the product you need to purchase.

The fact is, some suppliers really do not care or need to exert effort in retaining your company as a customer. They would rather lose the small amount of business your company gives them than spend the money or time it takes to meet requirements.

This does not mean your company should "give up" in trying to get the performance you need from this type of supplier. A couple of suggestions that may help you in dealing with this problem are:

1. Seek out other individuals within the supplier's organization who may be able to help you. Perhaps a meeting with the General Manager or President of that large company would be beneficial in an agreement being made.
2. Involve your customer purchasing activity, especially if it is your customer who has subjected you to this supplier. Explain to him/her that the supplier is uncooperative and not meeting requirements. Though the supplier thinks of your company as "small potatoes," they probably get a lot business through your customer.

Be creative and persistent when relying on suppliers to meet your company's requirements.

Index